WHEN GRAVITY BREAKS DOWN

BALUNGI FRANCIS

ALSO BY BALUNGI FRANCIS

Quantum Gravity in a Nutshell1

Fifty Formulas that Changed the World

Brief Solutions to the Big Problems in Physics, Astrophysics and cosmology

What is Real?:Space Time Singularities or Quantum Black Holes?Dark Matter or Planck Mass Particles? General Relativity or Quantum Gravity? Volume or Area Entropy Law?

Balungi's Guide to a Healthy Pregnancy

The origin of Gravity and the Laws of Physics: A new view on Gravity and the Cosmos

Copyright © Balungi Francis 2020
Copyright © Barungi Francis 2020
Copyright © Bill Stone Services 2020

Balungi Francis asserts the moral right to be identified as the author of this work.

All rights reserved. Apart from any fair dealing for the purposes of research or private study or critism or review, no part of this publication may be reproduced, distributed, or transmitted in any form or by any means, including photocopying, recording, or other electronic or mechanical methods, or by any information storage and retrieval system without the prior written permission of the publisher.

TABLE OF CONTENTS

Newton's Gravity Breaks Down 1

Theoretical Failures of Newton's Laws 4

Observational Conflicts 5

Einstein's General Relativity Breaks Down 8

Predictions of General Relativity 8

Failures of General Relativity 9

MOND Tries 13

MOND Success 16

MOND Failures 18

Gravitons are Promising 20

Difficulties and outstanding issues 22

Quantum Gravity Waits 24

String Theory Fails 29

Emergent Gravity Welcomed 33

I Proposed the Fifth Force 38

Gravity Fixed 43

Glossary 50

Bibliography 69

Acknowledgments 95
About the Author 96

PREFACE

In Newton's view, all objects exert a force that attracts other objects. That universal law of gravitation worked pretty well for predicting the motion of planets as well as objects on Earth and it's still used, for example, when making the calculations for a rocket launch. But Newton's view of gravity didn't work for some things, like Mercury's peculiar orbit around the sun. The orbits of planets shift over time, and Mercury's orbit shifted faster than Newton predicted.

Einstein's idea was that gravity is not a force, but it is really an effect caused by the curvature of space and time. Matter curves space-time in its vicinity, and this curvature in return affects how matter moves. This means that, according to Einstein, space and time are responsive. Although all its predictions have almost been confirmed by experiment, General relativity fails to explain details near space time singularities at the centre of Black holes and the mysterious dark matter. Which means Einstein equations cannot explain the motion of stars in galaxies and galaxy clusters.

MOND was proposed by Mordehai Milgrom in 1983. MOND explains the motion of stars in galaxies correctly without assuming Dark matter. Therefore MOND is an alternative to Newton's law of Universal gravitation. However the most serious problem facing Milgrom's law is that it cannot completely eliminate the need for dark matter in all astrophysical systems: galaxy clusters show a residual mass discrepancy even when analysed using MOND.

Most theorists believe that gravitons must exist, and that they could be candidates for Dark matter, because quantum theory has successfully explained every other force of nature. But not everyone agrees. No theory claiming to unify quantum theory with GR has been successfully verified, and this has raised suspicions that perhaps gravity isn't like any other force – in which case gravitons may not exist. Even if they do, finding them is another matter. Quantum theory predicts that as gravity has an effectively infinite range, the graviton must have an incredibly low mass. Studies of gravitational waves from colliding black holes suggest that the graviton must be at least a billion, billion, billion times lighter even than the electron.

The theory of quantum gravity is expected to be able to provide a satisfactory description of the microstructure of space time at the so called Planck scales, at which all fundamental constants of the ingredient theories, c (speed of light), h (Planck constant) and G (Newton's constant), come together to form units of mass, length and time. The search for the full theory of quantum gravity has been stymied by the fact that gravity's quantum properties never seem to manifest in actual experience. One of the issues with theories of quantum gravity is that their predictions are usually nearly impossible to experimentally test. This is the main reason why there exist so many competing theories and why we haven't been successful in understanding how it actually works.

One option for a solution to this conundrum is string theory, or the idea that everything we perceive as a particle or force is simply an excitation of a closed or open string, vibrating at specific but unique frequencies. One of the major criticisms of string theory has to do not with the theory so much as with theorists. The argument is that they are forming something of a "cult" of string theorists, who have bonded together to promote string theory above all alternatives. Not only that, the strings of string theory are stupendously small, thought to be somewhere around the Planck scale, a bare 10^{-34} meters across. That's far, far smaller than anything we can possibly hope to probe even with our most precise instruments. The strings are so small, in fact, that they appear to us to be point-like particles, such as electrons and photons and neutrons. We simply can't ever stare at a string directly.

Therefore emergent gravity or entropic gravity is a theory in modern physics that describes gravity as an entropic force, a force with macro-scale homogeneity but which is subject to quantum level disorder and not a fundamental interaction. The theory, based on string theory, black hole physics, and quantum information theory, describes gravity as an *emergent* phenomenon that springs from the quantum entanglement of small bits of spacetime information. As such, entropic gravity is said to abide by the second law of thermodynamics under which the entropy of a physical system tends to increase over time. The theory claims to be consistent with both the macro-level observations of Newtonian gravity as well as Einstein's theory of general relativity and its gravitational distortion of spacetime. Importantly, the theory also explains why galactic rotation curves differ from the profile expected with visible matter whithout invoking dark matter. The theory has been controversial within the physics community but has sparked research and experiments to test its validity. The problem is, if emergent gravity just reproduces General Relativity, there's no way to test the idea. What we need instead is a prediction from emergent gravity that deviates from General Relativity.

Finally when all is said and done, the fifth force is proposed and its first result is that it reproduces the MOND and Emergent gravity results from one single force equation and solves all gravitational problems leaving none untouched. Although it is accurate there is one problem; it can't explain the origin of gravity.

Gravity is then fixed by postulating that it is as a result of the Casimir effect due to vacuum polarizations with sounding experimental proof. This book will help you fix Gravity.

Newton's Gravity Breaks Down

.

Discovering a new law of nature is the acme of scientific achievement, and one granted to few. Those who succeed are assured of a place in the pantheon of science, first published in 1687, his book, "Mathematical Principles of Natural Philosophy," put forward his three laws of motion and his law of universal gravity. Sir Isaac Newton portrayed gravity as some kind of mysterious influence that allows masses to affect each other even through the vacuum of space. While declining to say exactly how this influence worked, Newton came up with a precise mathematical description of its effects, in the form of his celebrated "inverse-square law" of universal gravitation. Supposedly inspired by watching an apple fall in his mother's garden almost 350 years ago, Newton's law remained the best description of gravity until 1915, when Albert Einstein published his general theory of relativity, which gave the first detailed account of what gravity actually is.

In Newton's view, all objects exert a force that attracts other objects. That universal law of gravitation worked pretty well for predicting the motion of planets as well as objects on Earth and it's still used, for example, when making the calculations for a rocket launch.

But Newton's view of gravity didn't work for some things, like Mercury's peculiar orbit around the sun. The orbits of planets shift over time, and Mercury's orbit shifted faster than Newton predicted.

While Newton was able to formulate his law of gravity in his monumental work, he was deeply uncomfortable with the notion of "action at a distance" that his equations implied. In 1692, in his third letter to Bentley, he wrote:

"That one body may act upon another at a distance through a vacuum without the mediation of anything else, by and through which their action and force may be conveyed from one another, is to me so great an absurdity that, I believe, no man who has in philosophic matters a competent faculty of thinking could ever fall into it."

He never, in his words, "assigned the cause of this power". In all other cases, he used the phenomenon of motion to explain the origin of various forces acting on bodies, but in the case of gravity, he was unable to experimentally identify the motion that produces the force of gravity. Moreover, he refused to even offer a hypothesis as to the cause of this force on grounds that to do so was contrary to sound science. He lamented that "philosophers have hitherto attempted the search of nature in vain" for the source of the gravitational force, as he was convinced "by many reasons" that there were "causes hitherto unknown" that were fundamental to all the "phenomena of nature". These fundamental phenomena are still under investigation and, though hypotheses abound, the definitive answer has yet to be found.

"I have not yet been able to discover the cause of these properties of gravity from phenomena and I feign no hypotheses.... It is enough that gravity does really exist and acts according to the laws I have explained, and that it abundantly serves to account for all the motions of celestial bodies."

When Isaac Newton put forth his universal theory of gravitation in the 1680s, it was immediately recognized for what it was: the first incredibly successful, predicatively powerful scientific theory that described the one force ruling the largest scales of all. From objects freely falling here on Earth to the planets and celestial bodies orbiting in space, Newton's theory of gravity captured their trajectories spectacularly. When the new planet Uranus was discovered, the deviations in its orbit from Newton's predictions allowed a spectacular leap: the prediction of the existence, mass and position of another new world beyond it: Neptune. The very night the Berlin Observatory received the theoretical prediction of Urbain Le Verrier working 169 years after Newton's Principia they found our Solar System's 8th planet within one degree of its predicted position. And yet, Newton's laws were about to prove insufficient for what was to come.

The problem all started not at the outer reaches of the Solar System, but in the *innermost* regions: with the planet Mercury, orbiting closest to the Sun. Every planet orbits the Sun not in a perfect circle, but rather in an ellipse, as Kepler noticed nearly a full century before Newton. The orbits of Venus and Earth are very close to circular, but both Mercury and Mars are noticeably more elliptical, with their closest approach to the Sun differing significantly from their greatest distance.

Mercury, in particular, reaches a distance that's 46% greater at aphelion (its farthest point from the Sun) than at perihelion (its closest approach), as compared to just a difference of 3.4% from Earth. This doesn't have anything to do with the theory of gravity; this is merely the conditions which these planets formed under that led to these orbital properties. But the fact that these orbits aren't perfectly circular means we can study something interesting about them. If Kepler's laws were absolutely perfect, then a planet orbiting the Sun would return to the *exact same spot* with each and every orbit. When we reached perihelion one year, then if we counted out exactly one year, we'd expect to be at perihelion once again, and we'd expect the Earth to be in the same exact position in space relative to all the other stars and the Sun as it was the year before.

Pretty well, but not perfectly: Einstein showed that Newton's formula starts to break down when gravitational fields become very strong - for example, close to stars or black holes.

Einstein offered a different view of gravity, one that made sense of Mercury. Instead of exerting an attractive force, he reasoned that each object curves the fabric of space and time around them, forming a sort of well that other objects and even beams of light fall into. Think of the sun as a bowling ball on a mattress. It creates a depression that draws the planets close.

This new model solved the Mercury problem. It showed that the sun so curves space that it distorts the orbits of nearby bodies, including Mercury. In Einstein's view, Mercury might look like a marble forever circling the bottom of a drain.

Theoretical Failures of Newton's Laws

There is no immediate prospect of identifying the mediator of gravity. Attempts by physicists to identify the relationship between the gravitational force and other known fundamental forces are not yet resolved, although considerable headway has been made over the last 50 years.

Newton's theory of gravitation requires that the gravitational force be transmitted instantaneously. The significant propagation delay in gravity leads to unstable planetary and stellar orbits.

Observational Conflicts

Newton's theory does not fully explain the precession of the perihelion of the orbits of Planet Mercury. There is a 43 arcsecond per century discrepancy between the Newtonian calculation, which arises only from the gravitational attractions from the other planets, and the observed precession, made with advanced telescopes during the 19th century.

The predicted angular deflection of light rays by gravity that is calculated by using Newton's Theory is only one half of the deflection that is actually observed by astronomers.

In spiral galaxies, the orbiting of stars around their centers seems to strongly disobey Newton's law of universal gravitation.

During the 1970s, astronomers discovered something odd about the movement of stars in galaxies. Like the planets orbiting our sun, the stars should follow Newton's law of gravity, and travel ever more slowly the further out they are from the galactic centre. Yet beyond a certain distance, their speeds remained more or less constant - in flat contradiction of Newton's law.

Astronomers quickly proposed a solution: that there are huge amounts of invisible "dark matter" lurking in and around galaxies, whose gravitational pull invisibly affects the stars. But Prof Milgrom had a more radical proposal: that there is something wrong with the law of gravity itself. His calculations suggested that the anomalous motion of the stars could be explained if Newton's law breaks down for masses accelerating below a critical rate of around one ten-billionth of a metre per second per second.

For over 25 years Professor Mordehai Milgrom of the Weizmann Institute in Israel has been pursuing the possibility that both Newton and Einstein missed something when they devised their theories of this most ubiquitous of forces.

According to Einstein, mass warps the very fabric of space and time around it, rather like a cannonball sitting on a vast rubber sheet. This creates the illusion that objects moving past some mass are accelerated by a mysterious "force" emanating from it. In reality, they are just responding to the distortion of space and time - the effect of which is described in detail by Einstein's theory, and captured pretty well even by Newton's simple formula.

Since the early 1980s, Prof Milgrom has suspected there is another flaw in Newton's venerable formula - one which even Einstein failed to fix. And after decades of being ignored by the scientific establishment, there is mounting evidence that he is right.

Prof Milgrom's theory goes by the prosaic name of Modified Newtonian Dynamics or MOND, and is based the bizarre idea that Newton's law of gravity breaks down at low accelerations. And he means very low: around 100-billionth that generated by the Earth's gravity. Like Newton, Prof Milgrom was inspired by a simple observation - albeit a rather more esoteric one than the fall of an apple.

Newton once said that he saw himself as "only like a boy playing on the seashore... whilst the great ocean of truth lay all undiscovered before me"

Einstein's General Relativity Breaks Down

General relativity, the theory of gravity Albert Einstein published 100 years ago, is one of the most successful theories we have. It has passed every experimental test; every observation from astronomy is consistent with its predictions.

Einstein's idea was that gravity is not a force, but it is really an effect caused by the curvature of space and time. Matter curves space-time in its vicinity, and this curvature in return affects how matter moves. This means that, according to Einstein, space and time are responsive.

Predictions of General Relativity

It predicts that light rays bend around massive objects, like the sun, which we have observed. The same effect also gives rise to gravitational lensing, which we have also observed.

General Relativity further predicts that the universe should expand, which it does.

It predicts that time runs more slowly in gravitational potentials, which is correct.

General Relativity predicts black holes, and it predicts just how the black hole shadow looks, which is what we have observed.

It also predicts gravitational waves, which we have observed.

So, there is no doubt that General Relativity works extremely well. But we already know that it cannot ultimately be the correct theory for space and time. It is an approximation that works in many circumstances, but fails in others.

Failures of General Relativity

There are serious problems with local energy-momentum conservation in general relativity. It is well known that Einstein's theory does not assign a definite stress-energy tensor to the gravitational field. This property is extremely unsatisfactory, because one knows that all other fundamental interactions in nature actually do respect the principle of local conservation of energy-momentum. Essentially, the non-existance of a stress-energy tensor is a consequence of the purely geometrical interpretation of gravity as curvature of space-time.

Space-time singularities and event horizons are a consequence of general relativity, appearing in the solutions of the gravitational field. Although the "big bang" singularity and "black holes" have been a topic of intensive study in theoretical astrophysics, one can seriously doubt that such mathematical monsters should really represent physical objects. In fact, in order to predict black holes one has to extrapolate the theory of general relativity far beyond observationally known gravity strengths. In any other theory that we have, singularities are a sign that the theory breaks down and has to be replaced by a more fundamental theory. And we think the same has to be the case in General Relativity, where the more fundamental theory to replace it is quantum gravity.

Quantum mechanics can be said to be *the* cornerstone of modern physics. For every physical field theory it should be possible to formulate it as quantum field theory. Actually, it is generally accepted that the field theories of electromagnetism or gravitation are but an approximation, the "classical limit", of more fundamental underlying quantum field theories. It is also assumed that interaction theories have to be gauge theories. The possibility of formulating gravity as quantum field theory is essential in the context of the unification of all fundamental interactions. However, all attempts to find a consistent quantum gauge field theory of general relativity have failed. This indicates again that general relativity can hardly be an absolutely correct theory of gravitation

We know experimentally that particles have some strange quantum properties. They obey the uncertainty principle and they can do things like being in two places at once. Concretely, think about an electron going through a double slit. Quantum mechanics tells us that the particle goes through both slits. Now, electrons have a mass and masses generate a gravitational pull by bending space-time. This brings up the question, to which place does the gravitational pull go if the electron travels through both slits at the same time. You would expect the gravitational pull to also go to two places at the same time. But this cannot be the case in general relativity, because general relativity is not a quantum theory. To solve this problem, we have to understand the quantum properties of gravity. We need what physicists call a theory of quantum gravity. And since Einstein taught us that gravity is really about the curvature of space and time, what we need is a theory for the quantum properties of space and time.

The other reason we think gravity must be quantized is the trouble with information loss in black holes. If we combine quantum theory with general relativity but without quantizing gravity, then we find that black holes slowly shrink by emitting radiation. This was first derived by Stephen Hawking in the 1970s and so this black hole radiation is also called Hawking radiation.

Now, it seems that black holes can entirely vanish by emitting this radiation. Problem is, the radiation itself is entirely random and does not carry any information. So when a black hole is entirely gone and all you have left is the radiation, you do not know what formed the black hole. Such a process is fundamentally irreversible and therefore incompatible with quantum theory. It just does not fit together. A lot of physicists think that to solve this problem we need a theory of quantum gravity.

That hasn't stopped maverick scientists like Moffat from looking at alternatives to GR. The rotation of spiral galaxies inspired a particular modification to gravity that lingers like a fungus in the basement of astronomy: "modified Newtonian dynamics," or MOND. As the name suggests, it's a change to Newton's law of gravity rather than general relativity, and it does very well at describing the motion of stars and gas in spiral galaxies without the need for dark matter. However, MOND fails for some other types of galaxies, galaxy clusters, and—because it isn't compatible with relativity it cannot explain the "classic tests" of GR, much less the evolution of the universe as a whole.

John Moffat was also motivated to modify GR by the problem of dark matter, but is uninterested in reproducing MOND because of its observational failures. Instead, his modification of general relativity involves allowing the strength of gravity to vary slightly in space and time and changing the way gravity acts over long distances.

Few astrophysicists doubt that black holes exist: We know of a large number of very massive, very dense objects in the cosmos, for which the black hole hypothesis is the only one that fits. However, we have yet to "see" the event horizon, the boundary separating the exterior of a black hole from its interior—where nothing can escape back into the outside Universe. That's the goal of the Event Horizon Telescope (EHT), which is actually made of six observatories scattered around the world, observing the same objects in concert. Working together, they can create real images of whatever is right outside the black hole at the center of the Milky Way, a new frontier where GR's most exotic effects could be measured.

The ultimate arbiter of a theory, after all, is nature. If one of the dark matter experiments found particles with the right properties, then the motive to modify GR would diminish; if more and more experiments fail to find dark matter, then researchers are likely to pay more attention to alternative theories, perhaps even ones that are unorthodox or complex.

MOND Tries

MOND was proposed by Mordehai Milgrom in 1983. The basic premise of MOND is that while Newton's laws have been extensively tested in high-acceleration environments, they have not been verified for objects with extremely low acceleration, such as stars in the outer parts of galaxies. Several independent observations point to the fact that the visible mass in galaxies and galaxy clusters is insufficient to account for their dynamics, when analysed using Newton's laws.

While Newton's Laws predict that stellar rotation velocities should decrease with distance from the galactic centre, Rubin and collaborators found instead that they remain almost constant the rotation curves are said to be "flat". This observation necessitates at least one of the following:

1) There exists in galaxies large quantities of unseen matter which boosts the stars' velocities beyond what would be expected on the basis of the visible mass alone, or

2) Newton's Laws do not apply to galaxies. The former leads to the dark matter hypothesis; the latter leads to MOND.

In the disc galaxies most of the mass is at the centre of the galaxy, this means that if you want to calculate how a star moves far away from the centre it is a good approximation to only ask what is the gravitational pull that comes from the centre bulge of the galaxy. Einstein taught us that gravity is really due to the curvature of space and time but in many cases it is still quantitatively incorrect to describe gravity as a force, this is known as the Newtonian limit and is a good approximation as long as the pull of gravity is weak and objects move much slower than the speed of light. It is a bad approximation for example close by the horizon of a black hole but it is a good approximation for the dynamics of galaxies that we are looking at here. It is then not difficult to calculate the stable orbit of a star far away from the centre of a disc galaxy. For a star to remain on its orbit, the gravitational pull must be balanced by the centrifugal force, $\frac{mv^2}{R} = \frac{GMm}{R^2}$. You can solve this equation for the velocity of the star and this will give you the velocity that is necessary for a star to remain on a stable orbit, $v = \sqrt{\frac{GM}{R}}$. As you can see the velocity drops inversely with the square root of the distance to the centre. But this is not what we observe, what we observe instead is that the velocity continue to increase with distance from the galactic centre and then they become constant.

This is known as the flat rotation curve. This is not only the case for our own galaxy but it is the case for hundred of galaxies that have been observed. The curves don't always become perfectly constant sometimes they have rigorous lines but it is abundantly clear that these observations cannot be explained by the normal matter only.

Dark matter solves this problem by postulating that there is additional mass in galaxies distributed in a spherical halo. This has the effect of speeding up the stars because the gravitational pull is now stronger due to the mass from the dark matter halo. There is always a distribution of dark matter that will reproduce whatever velocity curve we observe.

In contrast to this, Modified Newtonian Dynamics (MOND) postulates that gravity works differently. In MOND, the gravitational potential is the logarithmic of the distance $\Phi = \left(\sqrt{GMa_o}\right) \ln\left(\frac{R}{GM}\right)$, and not as normally the inverse of the distance $\Phi = \frac{-GM}{R}$.

In MOND the gravitational force is then the derivation of the potential that is, the inverse of the distance $F = \frac{\sqrt{GMa_o}}{R}$, while normally it is the inverse of the square of the distance $F = \frac{GMm}{R^2}$. If you put this modified gravitational force into the force balance equation as before $\frac{\sqrt{GMa_o}}{R} = \frac{v^2}{R}$, you will see that the dependence on the distance cancels out and the velocity just becomes constant. Now of course you cannot just go and throw out the normal $\frac{1}{R^2}$ gravitational force law because we know that it works on the solar system. Therefore MOND postulates that the normal $\frac{1}{R^2}$ law crosses over into a $\frac{1}{R}$ law. This crossover happens not at a certain distance but it happens at a certain acceleration.

The New force law comes into play at low acceleration a_o, this acceleration where the crossover happens is a free parameter in MOND. You can determine the value of this pararmeter by just trying out which fits the data best. It turns out that the best fit value is closely related to the cosmological constant $a_o \approx \sqrt{\frac{\Lambda}{3}}$, why does that so? No one has any idea and it is the aim of this section to find out why.

MOND Success

In addition to demonstrating that rotation curves in MOND are flat, it provides a concrete relation between a galaxy's total baryonic mass (the sum of its mass in stars and gas) and its asymptotic rotation velocity. Observationally, this is known as the baryonic Tully Fisher relation and is found to conform quite closely to the MOND prediction.

Milgrom's law fully specifies the rotation curve of a galaxy given only the distribution of its baryonic mass. In particular, MOND predicts a far stronger correlation between features in the baryonic mass distribution and features in the rotation curve than does the dark matter hypothesis.

It predicts a specific relationship between the acceleration of a star at any radius from the centre of a galaxy and the amount of unseen (dark matter) mass within that radius that would be inferred in a Newtonian analysis. This is known as the "mass discrepancy-acceleration relation", and has been measured observationally.

One aspect of the MOND prediction is that the mass of the inferred dark matter go to zero when the stellar centripetal acceleration becomes greater than a_0, where MOND reverts to Newtonian mechanics. In dark matter hypothesis, it is a challenge to understand why this mass should correlate so closely with acceleration, and why there appears to be a critical acceleration above which dark matter is not required.

In MOND, all gravitationally bound objects with $a < a_0$ – regardless of their origin – should exhibit a mass discrepancy when analysed using Newtonian mechanics, and should lie on the BTFR. Under the dark matter hypothesis, objects formed from baryonic material ejected during the merger or tidal interaction of two galaxies are expected to be devoid of dark matter and hence show no mass discrepancy. Three objects unambiguously identified as Tidal Dwarf Galaxies appear to have mass discrepancies in close agreement with the MOND prediction.

Recent work has shown that many of the dwarf galaxies around the Milky Way and Andromeda are located preferentially in a single plane and have correlated motions. This suggests that they may have formed during a close encounter with another galaxy and hence be Tidal Dwarf Galaxies. If so, the presence of mass discrepancies in these systems constitutes further evidence for MOND.

By itself, Milgrom's law is not a complete and self-contained physical theory, but rather an ad-hoc empirically motivated variant of one of the several equations that constitute classical mechanics. Its status within a coherent non-relativistic hypothesis of MOND is akin to Kepler's third law within Newtonian mechanics; it provides a succinct description of observational facts, but must itself be explained by more fundamental concepts situated within the underlying hypothesis.

The majority of astrophysicists and cosmologists accept dark matter as the explanation for galactic rotation curves, and are committed to a dark matter solution of the missing-mass problem. MOND, by contrast, is actively studied by only a handful of researchers. The primary difference between supporters of ΛCDM and MOND is in the observations for which they demand a robust, quantitative explanation and those for which they are satisfied with a qualitative account, or are prepared to leave for future work.

MOND Failures

The most serious problem facing Milgrom's law is that it cannot completely eliminate the need for dark matter in all astrophysical systems: galaxy clusters show a residual mass discrepancy even when analysed using MOND.

The 2006 observation of a pair of colliding galaxy clusters known as the "Bullet Cluster", poses a significant challenge for all theories proposing a modified gravity solution to the missing mass problem, including MOND.

Several other studies have noted observational difficulties with MOND. For example, it has been claimed that MOND offers a poor fit to the velocity dispersion profile of globular clusters and the temperature profile of galaxy clusters, that different values of a_0 are required for agreement with different galaxies' rotation curves, and that MOND is naturally unsuited to forming the basis of a hypothesis of cosmology.

Furthermore, many versions of MOND predict that the speed of light be different from the speed of gravity, but in 2017 the speed of gravitational waves was measured to be equal to the speed of light.

Besides these observational issues, MOND and its generalizations are plagued by theoretical difficulties, several ad-hoc and inelegant additions to general relativity are required to create a hypothesis with a non-Newtonian non-relativistic limit, the plethora of different versions of the hypothesis offer diverging predictions in simple physical situations and thus make it difficult to test the framework conclusively, and some formulations (most prominently those based on modified inertia) have long suffered from poor compatibility with cherished physical principles such as conservation laws.

Gravitons are Promising

In the late 1600s, Isaac Newton devised the first serious theory of gravity. He described gravity as a field that could reach out across great distances and dictate the path of massive objects like the Earth. Newton's theory was stunningly effective, yet the nature of the gravitational field remained a mystery. In 1915, Albert Einstein's theory of general relativity gave theorists their first look "under the hood" of gravity. What we call gravity, Einstein argued, is actually the distortion of space and time. The Earth looks like it's rounding the Sun in an ellipse, but it's actually following a straight line through warped spacetime.

Einstein's theory of gravity is very good at explaining the behavior of large objects. But just a few years later, physicists opened up the world of the ultra-small, revealing that the other fundamental forces are due to the exchange of specialized force-carrying particles: photons convey electromagnetism, the strong nuclear force is transmitted by gluons and the weak nuclear force is imparted by the movement of the W and Z bosons. Is gravity due to the same kind of particle exchange?

The graviton must have zero mass. Like massless photons, gravitons should travel at the speed of light. General relativity also gives us some insight into the nature of gravitons. In general relativity, the distribution of mass and energy in the universe is described by a four-by-four matrix that mathematicians call a tensor of rank two. This is important because if the tensor is the source of gravitation, you can show that the graviton must be a particle with a quantum mechanical spin of two. Another nice fallout of this correspondence is that the graviton is the only possible massless, spin two particle. If you observe a massless, spin two particle, you have found the graviton.

Gravitons are at the heart of arguably the biggest challenge in theoretical physics: the search for the 'theory of everything' – a set of equations describing all of the forces and particles in the Universe.

Most theorists believe that gravitons must exist, because quantum theory has successfully explained every other force of nature. But not everyone agrees. No theory claiming to unify quantum theory with GR has been successfully verified, and this has raised suspicions that perhaps gravity isn't like any other force – in which case gravitons may not exist.

Even if they do, finding them is another matter. Quantum theory predicts that as gravity has an effectively infinite range, the graviton must have an incredibly low mass. Studies of gravitational waves from colliding black holes suggest that the graviton must be at least a billion, billion, billion times lighter even than the electron.

The problem with searching for gravitons is that gravity is incredibly weak. For instance, the electromagnetic force between an electron and a proton in a hydrogen atom is 10^{39} times larger than the gravitational force between the same two particles.

Individual gravitons interact very feebly, and we are only held to the planet because the Earth emits so many of them. Because a single graviton is so weak, it is impossible for us to directly detect individual classical gravitons.

However, there are new and innovative ideas about gravity in which other forms of gravitons might exist. Some of these exotic gravitons might be detectable, but they require significant modifications to our understanding of our universe. This is where things get a bit mind-bending.

Gravity is the one known fundamental force that has resisted study in the quantum realm and finding gravitons of any kind would be a huge step forward in our understanding of the phenomenon. Devising a successful theory of quantum gravity is one of the hottest goals of modern physics and ongoing experimental searches for gravitons will play a central role.

Gravity is also by far the feeblest fundamental force in nature. This means that any graviton detector must be incredibly massive and placed near a powerful source of gravitons. Calculations suggests that even a detector with the mass of Jupiter orbiting a bizarre object like a neutron star (a potential strong source of gravitons) would struggle to find anything.

Difficulties and outstanding issues

Attempts to extend the Standard Model or other quantum field theories by adding gravitons run into serious theoretical difficulties at energies close to or above the Planck scale. This is because of infinities arising due to quantum effects; technically, gravitation is not renormalizable. Since classical general relativity and quantum mechanics seem to be incompatible at such energies, from a theoretical point of view, this situation is not tenable.

The graviton remains hypothetical, however, because at the moment, it's impossible to detect. Although gravity on a planetary scale is strong, on a small scales it can be very feeble. So much so that when a magnet attracts a paperclip, it pulls against the gravitational force of the entire planet, and still wins. This means that a single graviton if it exists is very, very weak.

There are many things about quantum physics that we don't understand and understanding particles and the laws that govern them can help us wield the powers that quantum phenomena hypothetically possess. Proving the existence of a particle that would help make sense of it all is a dream, and remains that as of now.

As it stands, we are far from definitively proving it exists. As Fermilab senior physicist Don Lincoln wrote in a post: "Gravitons are a theoretically reputable idea, but are not proven. So if you hear someone say that 'gravitons are particles that generate the gravitational force,' keep in mind that this is a reasonable statement, but by no means is it universally accepted. It will be a long time before gravitons are considered part of the established subatomic pantheon."

Quantum Gravity Waits

The development of a quantum theory of gravity began in 1899 with Max Planck's formulation of "Planck scales" of mass, time, and length. During this period, the theories of quantum mechanics, quantum field theory and general relativity had not yet been developed. This means that Planck himself had no idea about what he had just developed-behind the Black board. Planck was not aware of quantum gravity and what it would mean for physicists. But he had just coined in formula one of the starting point for the holy grail of physics.

After P.Bridgman's disapproval of Planck's units in 1922, Albert Einstein having published the General Relativity theory, a few months after its publication he noted that "to the intra-atomic movement of electrons, atoms would have to radiate not only electromagnetic but also gravitational energy if only in tiny amounts, as this is hardly true in nature, it appears that quantum theory would have to modify not only Maxwellian electrodynamics, but also the new theory of gravitation". This showed Einstein's interest in the unification of Planck's quantum theory with his newly developed theory of Gravitation.

Then in 1933 came Bronstein's cGh-plan as we know it today. In his plan he argued a need for Quantum Gravity. In his own words he stated: "After the relativistic quantum theory is created, the task will be to develop the next part of our scheme that is, to unify quantum theory (h), special relativity (c) and the theory of gravitation (G) into a single theory". Bronstein figured out how to describe gravity in terms of quantized particles, now called gravitons, but only when the force of gravity is weak — that is (in general relativity), when the space-time fabric is so weakly curved that it can be approximated as flat. When gravity is strong, "the situation is quite different," he wrote. "Without a deep revision of classical notions, it seems hardly possible to extend the quantum theory of gravity also to this domain."

Thus the theory of quantum gravity is expected to be able to provide a satisfactory description of the microstructure of space time at the so called Planck scales, at which all fundamental constants of the ingredient theories, c (speed of light), h (Planck constant) and G (Newton's constant), come together to form units of mass, length and time.

Therefore the need for the theory of quantum gravity is crucial in understanding nature, from the smallest to the biggest particle ever known in the universe. For example, "we can describe the behavior of flowing water with the long- known classical theory of hydrodynamics, but if we advance to smaller and smaller scales and eventually come across individual atoms, it no longer applies. Then we need quantum physics just as a liquid consists of atoms" Daniel Oriti in this case imagines space to be made up of tiny cells or atoms of space and a new theory of quantum gravity is required to describe them fully.

The problem is that even though gravity keeps us stuck to the ground and generally acts as a force, general relativity suggests it's something more the shape of space itself. Other quantum theories treat space as a flat backdrop for measuring how far and fast particles fly. Ignoring the curvature of space for particles works because gravity is so much weaker than the other forces that space looks flat when zoomed in on something as small as an electron. The effects of gravity and the curvature of space are relatively obvious at more zoomed-out levels, like planets and stars. But when physicists try to calculate the curvature of space around an electron, slight as it may be, the math becomes impossible.

One theory, known as loop quantum gravity, aims to resolve the conflict between particles and space-time by breaking up space and time into little bits an ultimate resolution beyond which no zooming can take place.

String theory, another popular framework, takes a different approach and swaps out particles for fiber-like strings, which behave better mathematically than their point-like counterparts. This simple change has complex consequences, but one nice feature is that gravity just falls out of the math. Even if Einstein and his contemporaries had never developed general relativity, Engelhardt said, physicists would have stumbled upon it later through string theory. "I find that pretty miraculous," she said.

The search for the full theory of quantum gravity has been stymied by the fact that gravity's quantum properties never seem to manifest in actual experience. Physicists never get to see how Einstein's description of the smooth space-time continuum, or Bronstein's quantum approximation of it when it's weakly curved, goes wrong.
The problem is gravity's extreme weakness. Whereas the quantized particles that convey the strong, weak and electromagnetic forces are so powerful that they tightly bind matter into atoms, and can be studied in tabletop experiments, gravitons are individually so weak that laboratories have no hope of detecting them. To detect a graviton with high probability, a particle detector would have to be so huge and massive that it would collapse into a black hole. This weakness is why it takes an astronomical

accumulation of mass to gravitationally influence other massive bodies, and why we only see gravity writ large.

Not only that, but the universe appears to be governed by a kind of cosmic censorship Regions of extreme gravity where space-time curves so sharply that Einstein's equations malfunction and the true, quantum nature of gravity and space-time must be revealed always hide behind the horizons of black holes.

A crucial first step in this quest to know whether gravity is quantum is to detect the long-postulated elementary particle of gravity, the graviton. In search of the graviton, physicists are now turning to experiments involving microscopic superconductors, free-falling crystals and the afterglow of the big bang.

Quantum mechanics suggests everything is made of quanta, or packets of energy, that can behave like both a particle and a wave for instance, quanta of light are called photons. Detecting gravitons, the hypothetical quanta of gravity, would prove gravity is quantum. The problem is that gravity is extraordinarily weak. To directly observe the minuscule effects a graviton would have on matter, a graviton detector would have to be so massive that it collapses on itself to form a black hole.

"One of the issues with theories of quantum gravity is that their predictions are usually nearly impossible to experimentally test," says quantum physicist Richard Norte of Delft University of Technology in the Netherlands. "This is the main reason why there exist so many competing theories and why we haven't been successful in understanding how it actually works."

String Theory Fails

Newtonian gravity superseded Kepler's laws because of its additional predictive power, combining terrestrial and celestial mechanics. Even Einstein's relativity, both special and general, came about because of the failures of Newtonian mechanics to account for behavior close to the speed of light and in strong gravitational fields. It took observations well beyond what was capable of in Newton's time, such as the measurements of the lifetimes of particles produced in radioactive decays and the orbit of Mercury around the Sun over the course of centuries. The continued gathering of data in new regimes, at higher precision and over longer timescales allowed us to see the cracks in the scientific theories, as well as where the potential to expand beyond them were.

Now, we come to the present day. Einstein's general relativity is still our leading theory of gravity, having passed every experimental and observational test tossed its way, from gravitational lensing to relativistic frame dragging to the decay of binary pulsar orbits, while three other fundamental forces electromagnetism and the strong and weak nuclear forces are described by quantum field theories. These two classes of theories are fundamentally incompatible and incomplete on their own, and indicate that there is more to the Universe than we currently understand.

One option for a solution to this conundrum is string theory, or the idea that everything we perceive as a particle or force is simply an excitation of a closed or open string, vibrating at specific but unique frequencies.

String theory has been the darling of the theoretical physics community for decades. Brilliant theoretical physicists tell us that this theory is the best answer to the hardest problem that their field has ever attacked.

One of the major criticisms of string theory has to do not with the theory so much as with theorists. The argument is that they are forming something of a "cult" of string theorists, who have bonded together to promote string theory above all alternatives.

String theory is the most popular approach to a theory of quantum gravity, but that very phrase most popular is exactly the problem in the eyes of some. In physics, who cares (or who *should* care) how popular a theory is?

In fact, some critics believe that string theory is little more than a cult of personality. The practitioners of this arcane art have long ago foregone the regular practice of science, and now bask in the glory of seer-like authority figures like Edward Witten, Leonard Susskind, and Joe Polchinski, whose words can no more be wrong than the sun can stop shining. First, though, we have to examine why string theory is so hard to test. There are two reasons.

The strings of string theory are stupendously small, thought to be somewhere around the Planck scale, a bare 10^{-34} meters across. That's far, far smaller than anything we can possibly hope to probe even with our most precise instruments. The strings are so small, in fact, that they appear to us to be point-like particles, such as electrons and photons and neutrons. We simply can't ever stare at a string directly.

Related to that smallness is the energy scale needed to probe the regimes where string theory actually matters. As of today, we have two different approaches for explaining the four forces of nature. On one hand, we have the techniques of quantum field theory, which provide a microscopic description of electromagnetism and the two nuclear forces. And on the other we have general relativity, which allows us to understand gravity as the bending and warping of spacetime.

For all cases that we can directly examine, using one or the other is just fine. String theory only comes into play when we try to combine all four forces with a single description, which only really matters at the very highest energy scales so high that we could never, ever build a machine to reach such heights.

But even if we could devise a particle collider to directly probe the energies of quantum gravity, we couldn't test string theory, because as of yet string theory isn't complete. It doesn't exist. We only have approximations that we hope come close to the actual theory, but we have no idea how right (or wrong) we are. So string theory isn't even up to the task of making predictions that we could compare to hypothetical experiments.

One suggestion put forth by the string theorists is another kind of theoretical string: the cosmic string. Cosmic strings are universe-spanning defects in spacetime, leftover from the earliest moments of the Big Bang, and they're a pretty generic prediction of the physics of those epochs of the universe.

To date, no cosmic strings have been found in our universe. Still, the search is on. If we found a cosmic string, it wouldn't necessarily validate string theory there would be a lot more work needed to be done, both theoretically and observationally, to tell apart the string theory prediction from the crack-in-spacetime version.

Still, we might be able to pick up some interesting clues, and one of those clues is supersymmetry. Within string theory, supersymmetry allows the strings to describe not just the forces of nature but also the building blocks, giving that theory the power to truly be a theory of everything. If we found evidence for supersymmetry, it wouldn't prove string theory, but it would be a major stepping stone.

We haven't found any evidence for supersymmetry. Will we one day have evidence for even one of the underpinnings or side predictions of string theory? It's impossible to say. A lot of hopes were pinned on supersymmetry, which has so far failed to deliver, and questions remain about whether it's worth it to build even-larger colliders to try pushing harder on supersymmetry, or if we should just give up and try something else.

Emergent Gravity Welcomed

Gravity is one of the four fundamental forces of nature, which means it's not derived from anything else it just is. At least, that's according to our presently accepted theories. But this may be about to change.

It is 20 years since Ted Jacobson demonstrated that General Relativity resembles thermodynamics, which is a framework to describe how very large numbers of individual, constituent particles behave. Since then, physicists have tried to figure out whether this similarity is a formal coincidence or hints at a deeper truth: that space-time is made of small elements whose collective motion gives rise to the force we call gravity. In this case, gravity would not be a truly fundamental phenomenon, but an emergent one.

Therefore emergent gravity also entropic gravity, is a theory in modern physics that describes gravity as an entropic force, a force with macro-scale homogeneity but which is subject to quantum level disorder and not a fundamental interaction. The theory, based on string theory, black hole physics, and quantum information theory, describes gravity as an *emergent* phenomenon that springs from the quantum entanglement of small bits of spacetime information. As such, entropic gravity is said to abide by the second law of thermodynamics under which the entropy of a physical system tends to increase over time.

Entropic gravity provides the underlying framework to explain MOND, which holds that at a gravitational acceleration threshold of approximately 1.2×10^{-10} m/s^2, gravitational strength begins to vary *inversely* (linearly) with distance from a mass rather than the normal inverse square law of the distance.

The theory claims to be consistent with both the macro-level observations of Newtonian gravity as well as Einstein's theory of general relativity and its gravitational distortion of spacetime. Importantly, the theory also explains why galactic rotation curves differ from the profile expected with visible matter whithout invoking dark matter.

The theory has been controversial within the physics community but has sparked research and experiments to test its validity.

The problem is, if emergent gravity just reproduces General Relativity, there's no way to test the idea. What we need instead is a prediction from emergent gravity that deviates from General Relativity.

Such a prediction was made by Erik Verlinde. Verlinde pointed out that emergent gravity in a universe with a positive cosmological constant like the one we live in would only approximately reproduce General Relativity. The microscopic constituents of space-time, Verlinde claims, also react to the presence of matter in a way that General Relativity does not capture: they push inwards on matter. This creates an effect similar to that ascribed to particle dark matter, which pulls normal matter in by its gravitational attraction.

Verlinde's idea is interesting and solves two problems that had plagued previous attempts at emergent gravity.

First, he conjectures that the deviations from General Relativity come about because the microscopic constituents of space-time have an additional type of entropy. In the thermodynamic formulation of gravity, the entropy – that is the number of possible microscopic configurations – which a volume can maximally have is proportional to the surface area of that volume. This is also often referred to as a "holographic" entropy because it demonstrates that all what happens inside the volume can entirely be encoded on its surface. The additional entropy that Verlinde introduces instead grows with the volume itself.

The modification to General Relativity then comes about because matter – so the conjecture goes – reduces the new, volume-scaling entropy in its environment. The entropy decrease leads to a decrease in volume which, in turn creates a force pushing inwards on the matter. This force, Verlinde shows, is similar to the force normally attributed to dark matter – which pulls in normal matter from its additional gravitational mass.

However, the new entropy that Verlinde introduces can't become less than zero. Therefore, once the additional entropy is entirely depleted, one is left with only the usual, holographic entropy and gets back ordinary General Relativity. This happens in systems with a comparably high average density, such as solar systems. On galactic scales however, the modification to General Relativity becomes noticeable, and manifests itself as apparent dark matter. This solves a serious problem with many modifications of gravity which usually work well on galactic scales but not on solar system scales.

Second, Verlinde's idea explains a previously noted numeric coincidence. In modified gravity scenarios, the departure from General Relativity becomes relevant at a particular acceleration scale. That scale turns out to be similar – on the same order of magnitude – to the temperature of de-Sitter space, which is proportional to the (square root of the) cosmological constant. In the new emergent gravity model, this relation follows because the apparent dark matter is, in fact, related to the cosmological constant.

The real challenge for emergent gravity, I think, is not galactic rotation curves. That is the one domain where we already know that modified gravity at last some variants thereof work well. The real challenge is to also explain structure formation in the early universe, or any gravitational phenomena on larger (tens of millions of light years or more) scales.

One of the key differences between Verlinde's approach and that of others is the space in which he works. Our Universe is expanding, and that expansion is accelerating. This is called de Sitter space, and it comes with a positive cosmological constant driving the acceleration. But it's easier to do the mathematics in a Universe with a slowing rate of expansion and a negative cosmological constant (called anti-de Sitter space).

According to Hossenfelder, emergent gravity by itself isn't especially useful. "In the end, we already know general relativity, so you can say 'well what's the point?' What is the point in showing that [general relativity] can be re-expressed as a different theory? It's still the same outcome," she said. "Maybe it's a different way to calculate things. That's nice, but not really insightful."

While this re-casting of Einstein's equations as a consequence of thermodynamics isn't useful on its own, it is the foundation on which the rest of Verlinde's work is based.

I Proposed the Fifth Force

In the disc galaxies most of the mass is at the centre of the galaxy, this means that if you want to calculate how a star moves far away from the centre it is a good approximation to only ask what is the gravitational pull that comes from the centre bulge of the galaxy.

Einstein taught us that gravity is really due to the curvature of space and time but in many cases it is still quantitatively incorrect to describe gravity as a force, this is known as the Newtonian limit and is a good approximation as long as the pull of gravity is weak and objects move much slower than the speed of light. It is a bad approximation for example close by the horizon of a black hole but it is a good approximation for the dynamics of galaxies that we are looking at here. It is then not difficult to calculate the stable orbit of a star far away from the centre of a disc galaxy.

For a star to remain on its orbit, the gravitational pull must be balanced by the centrifugal force,

$$\frac{mv^2}{r} = \frac{GMm}{r^2}.$$

You can solve this equation for the velocity of the star and this will give you the velocity that is necessary for a star to remain on a stable orbit,

$$v = \sqrt{\frac{GM}{r}}.$$

As you can see the velocity drops inversely with the square root of the distance to the centre. But this is not what we observe, what we observe instead is that the velocity continue to increase with distance from the galactic centre and then they become constant.

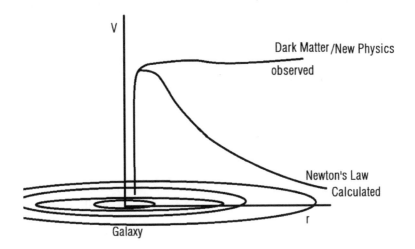

This is known as the flat rotation curve. This is not only the case for our own galaxy but it is the case for hundred of galaxies that have been observed. The curves don't always become perfectly constant sometimes they have rigorous lines but it is abundantly clear that these observations cannot be explained by the normal matter only.

Dark matter solves this problem by postulating that there is additional mass in galaxies distributed in a spherical halo. This has the effect of speeding up the stars because the gravitational pull is now stronger due to the mass from the dark matter halo. There is always a distribution of dark matter that will reproduce whatever velocity curve we observe.

This invisible and undetected matter removes any need to modify Newton's and Einstein's gravitational theories.

Invoking dark matter is a less radical, less scary alternative for most physicists than inventing a new theory of gravity.

If dark matter is not detected and does not exist, then Einstein's and Newton's gravity theories must be modified. Can this be done successfully? Yes! My Fifth force can explain the astrophysical, astronomical and cosmological data without dark matter as given below.

$$F = \sqrt{\alpha \hbar c \rho}$$

Where α is the coupling constant, $\hbar$ is the reduced Planck constant, c is the constant speed of light and ρ is the energy density.

The energy density is the force on particles moving in the vacuum per unit area,

$$\rho = \frac{F_o}{4\pi r^2}$$

From Newton's second law

$$F_o = m a_o$$

In contrast to Dark matter, our fifth force postulates that gravity works differently. The gravitational force is the inverse of the distance

$$F = \frac{1}{r}\sqrt{\alpha \hbar c m a_o}$$

While normally it is the inverse of the square of the distance

$$F = \frac{GMm}{r^2}$$

If you put this fifth force into the force balance equation as before

$$\frac{mv^2}{r} = \frac{1}{r}\sqrt{\alpha \hbar c m \, a_o}$$

you will see that the dependence on the distance cancels out and the velocity just becomes constant. Finally one inserts the gravitational coupling constant

$$\alpha = \frac{GMm}{\hbar c}$$

and one obtains the familiar relation

$$v^4 = a_o GM$$

We have recovered the Tully-Fisher relation, practically from first principles!

This Fifth force is an alternative to the hypothesis of dark matter in terms of explaining why galaxies do not appear to obey the currently understood laws of physics.

It is an alternative to Entropic gravity, MOND, Quantum gravity and General relativity.

Now of course you cannot just go and throw out the normal

$$\frac{1}{r^2}$$

gravitational force law because we know that it works on the solar system. Therefore the normal $\frac{1}{r^2}$ law crosses over into a $\frac{1}{r}$ law. This crossover happens not at a certain distance but it happens at certain acceleration. The New force law comes into play at low acceleration $a_o = 1.2 \times 10^{-10} ms^{-2}$, this acceleration where the crossover happens is a free parameter in MOND. You can determine the value of this pararmeter by just trying out which fits the data best. It turns out that the best fit value is closely related to the cosmological constant

$$a_o \approx \sqrt{\frac{\Lambda}{3}}$$

We anticipate that this fifth force will modify how stars collapse and the nature of black holes.

Gravity Fixed

In 2015 theoretical physicist James Quach, suggested a way to detect gravitons by taking advantage of their quantum nature. Quantum mechanics suggests the universe is inherently fuzzy for instance, one can never absolutely know a particle's position and momentum at the same time. One consequence of this uncertainty is that a vacuum is never completely empty, but instead buzzes with a "quantum foam" of so-called virtual particles that constantly pop in and out of existence. These ghostly entities may be any kind of quanta, including gravitons.

Decades ago, scientists found that virtual particles can generate detectable forces. For example, the Casimir effect is the attraction or repulsion seen between two mirrors placed close together in vacuum. These reflective surfaces move due to the force generated by virtual photons winking in and out of existence. Previous research suggested that superconductors might reflect gravitons more strongly than normal matter, so Quach calculated that looking for interactions between two thin superconducting sheets in vacuum could reveal a gravitational Casimir effect. The resulting force could be roughly 10 times stronger than that expected from the standard virtual-photon-based Casimir effect.

Recently, Norte and his colleagues developed a microchip to perform this experiment. This chip held two microscopic aluminum-coated plates that were cooled almost to absolute zero so that they became superconducting. One plate was attached to a movable mirror, and a laser was fired at that mirror. If the plates moved because of a gravitational Casimir effect, the frequency of light reflecting off the mirror would measurably shift. However the scientists failed to see any gravitational Casimir effect. This null result does not necessarily rule out the existence of gravitons and thus gravity's quantum nature. Rather, it may simply mean that gravitons do not interact with superconductors as strongly as prior work estimated, says quantum physicist and Nobel laureate Frank Wilczek of the Massachusetts Institute of Technology, who did not participate in this study and was unsurprised by its null results. Even so, Quach says, this "was a courageous attempt to detect gravitons."

Although Norte's microchip did not discover whether gravity is quantum, other scientists are pursuing a variety of approaches to find gravitational quantum effects. For example, in 2017 two independent studies suggested that if gravity is quantum it could generate a link known as "entanglement" between particles, so that one particle instantaneously influences another no matter where either is located in the cosmos. A table top experiment using laser beams and microscopic diamonds might help search for such gravity-based entanglement. The crystals would be kept in a vacuum to avoid collisions with atoms, so they would interact with one another through gravity alone. Scientists would let these diamonds fall at the same time, and if gravity is quantum the gravitational pull each crystal exerts on the other could entangle them together.

The researchers would seek out entanglement by shining lasers into each diamond's heart after the drop. If particles in the crystals' centers spin one way, they would fluoresce, but they would not if they spin the other way. If the spins in both crystals are in sync more often than chance would predict, this would suggest entanglement. "Experimentalists all over the world are curious to take the challenge up," says quantum gravity researcher Anupam Mazumdar of the University of Groningen in the Netherlands, co-author of one of the entanglement studies.

No, we have not failed. At least we know that the space between the plates is not completely empty. The virtual electromagnetic waves (photons) between the plates move at a constant speed of light c and so we can start from there.

The Casimir force between the plates was computed and has been verified by experiment to be

$$F = \frac{4\pi \hbar c}{A}$$

Where c is the speed of light and A is the surface area of the plates.

Let us replace the plates with an electron of mass m and charge e. let the surface area of the electron be A. Now let us calculate the force felt by the electron from first principles.

Let us assume that the electron is placed in the vacuum "empty space", a sea of virtual photons popping in and out of space at the speed of light c.

According to the Heisenberg Uncertainty principle, empty space isn't completely empty but consists of an indeterminate state of fluctuating fields and particles.

Even if you remove all the particles and radiation from a region of space, space still won't be empty. It will consist of virtual pairs of particles and antiparticles.

It is known that, when light propagates through an "empty" region (Vacuum), if space is perfectly empty, it should move through that space unimpeded, without bending, slowing or breaking into multiple wavelengths. Applying an external magnetic field doesn't change this, as photons, with their oscillatory electric and magnetic fields don't bend in a magnetic field. Even when your space is filled with particle antiparticle pairs, this effect doesn't change. But if you apply a **strong magnetic field** to a space filled with particle antiparticle pairs, suddenly a real, observable effect arises.

The speed of the virtual photons in a vacuum is related to the electric E of the electron and the external magnetic field B on the electron by

$$c = \frac{E}{B}$$

The force felt by an electron due to vacuum polarization then becomes,

$$F = \frac{4\pi\hbar E}{BA}$$

The above given force will reduce to the familiar gravitational force felt by an electron when the following conditions become true:

1) When the characteristic value of the electric field built from the electron mass is

$$E = \frac{m^2 ec^2}{4\pi\varepsilon_o \hbar^2} = 9.667 \times 10^{15}\, \text{N/C}$$

2) When the characteristic value of the external Magnetic field through the surface area of the electron is

$$B = \frac{ec^2}{4\pi\varepsilon_o G\hbar} = 1.8423 \times 10^{48}\, T$$

3) When the surface area of the electron is

$$A = 4\pi r^2$$

Where r is the radius of the electron

When all the above given values have been put into consideration, one obtains the familiar law

$$F = G\frac{m^2}{r^2}$$

We have recovered Newton's law of gravitation, practically form first principles. Following the above derivation carefully, it implies that gravity is a force resulting from the quantum vacuum polarizations due to an existence of an external strong magnetic field. This is proof that gravity is indeed a quantum force that can be explained from the Casimir effect.

From the above given derivation, we notice that the gravitational force on an electron falls off as the square of its radius computed from its surface area. The effects of the given derivation can be experimentally verified near Neutron stars and Magnetars with strong magnetic fields. When the given values of the electric and magnetic fields are observed with high tech experiments then it will prove without doubt that the force felt by the electron in the vaccum is indeed a quantum force.

The above given theoretical model can be verified experimentally from the effect known as Vacuum Birefringence, occurring when charged particles, get yanked in opposite directions by strong magnetic field lines.

The effect of this vacuum birefringence gets stronger very quickly as the magnetic field strength increases, as the square of the field strength. Even though the effect is small, we have place in the universe where the magnetic field strength get large enough to make these effects relevant. This place is the Neutron star and Magnetars with strong magnetic fields.

The outer 10% of a neutron star consists mostly of protons, light nuclei, and electrons, which can stably exist without being crushed at the neutron star's surface.

Neutron stars rotate extremely rapidly, frequently in excess of the 10% the speed of light, meaning that these charged particles on the outskirts of the neutron star are always in motion, which necessitates the production of both electric currents and induced magnetic fields. These are the fields we should be looking for if we want to observe vacuum birefringence, and its effects on the polarization of light.

All the light that's emitted must pass through the strong magnetic field around the neutron star on its way to our eyes, telescope and detectors, if the magnetized space that it passes through exhibits the expected vacuum birefringence effect, that light should all be polarized, with a common direction of polarization for all the photons.

Glossary

Absolute space and time—the Newtonian concepts of space and time, in which space is independent of the material bodies within it, and time flows at the same rate throughout the universe without regard to the locations of different observers and their experience of "now."

Acceleration—the rate at which the speed or velocity of a body changes.

Accelerating universe—the discovery in 1998, through data from very distant supernovae, that the expansion of the universe in the wake of the big bang is not slowing down, but is actually speeding up at this point in its history; groups of astronomers in California and Australia independently discovered that the light from the supernovae appears dimmer than would be expected if the universe were slowing down.

Action—the mathematical expression used to describe a physical system by requiring only the knowledge of the initial and final states of the system; the values of the physical variables at all intermediate states are determined by minimizing the action.

Anthropic principle—the idea that our existence in the universe imposes constraints on its properties; an extreme version claims that we owe our existence to this principle.

Asymptotic freedom (or safety)—a property of quantum field theory in which the strength of the coupling between elementary particles vanishes with increasing energy and/or decreasing distance, such that the elementary particles approach free particles with no external forces acting on them; moreover for decreasing energy and/or increasing distance between the particles, the strength of the particle force increases indefinitely.

Baryon—a subatomic particle composed of three quarks, such as the proton and neutron.
Big bang theory—the theory that the universe began with a violent explosion of spacetime, and that matter and energy originated from an infinitely small and dense point.
Big crunch—similar to the big bang, this idea postulates an end to the universe in a singularity.
Binary stars—a common astrophysical system in which two stars rotate around each other; also called a "double star."
Blackbody—a physical system that absorbs all radiation that hits it, and emits characteristic radiation energy depending upon temperature; the concept of blackbodies is useful, among other things, in learning the temperature of stars.
Black hole—created when a dying star collapses to a singular point, concealed by an "event horizon;" the black hole is so dense and has such strong gravity that nothing, including light, can escape it; black holes are predicted by general relativity, and though they cannot be "seen," several have been inferred from astronomical observations of binary stars and massive collapsed stars at the centers of galaxies.
Boson—a particle with integer spin, such as photons, mesons, and gravitons, which carries the forces between fermions.
Brane—shortened from "membrane," a higher-dimensional extension of a onedimensional string.
Cassini spacecraft—NASA mission to Saturn, launched in 1997, that in addition to making detailed studies of Saturn and its moons, determined a bound on the variations of Newton's gravitational constant with time.

Causality—the concept that every event has in its past events that caused it, but no event can play a role in causing events in its past.

Classical theory—a physical theory, such as Newton's gravity theory or Einstein's general relativity, that is concerned with the macroscopic universe, as opposed to theories concerning events at the submicroscopic level such as quantum mechanics and the standard model of particle physics.

Copernican revolution—the paradigm shift begun by Nicolaus Copernicus in the early sixteenth century, when he identified the sun, rather than the Earth, as the center of the known universe.

Cosmic microwave background (CMB)—the first significant evidence for the big bang theory; initially found in 1964 and studied further by NASA teams in 1989 and the early 2000s, the CMB is a smooth signature of microwaves everywhere in the sky, representing the "afterglow" of the big bang: Infrared light produced about 400,000 years after the big bang had redshifted through the stretching of spacetime during fourteen billion years of expansion to the microwave part of the electromagnetic spectrum, revealing a great deal of information about the early universe.

Cosmological constant—a mathematical term that Einstein inserted into his gravity field equations in 1917 to keep the universe static and eternal; although he later regretted this and called it his "biggest blunder," cosmologists today still use the
cosmological constant, and some equate it with the mysterious dark energy.

Coupling constant—a term that indicates the strength of an interaction between particles or fields; electric charge and Newton's gravitational constant are coupling constants.

Crystalline spheres—concentric transparent spheres in ancient Greek cosmology that held the moon, sun, planets, and stars in place and made them revolve around the Earth; they were part of the western conception of the universe until the Renaissance.

Curvature—the deviation from a Euclidean spacetime due to the warping of the geometry by massive bodies.

Dark energy—a mysterious form of energy that has been associated with negative pressure vacuum energy and Einstein's cosmological constant; it is hypothesized to explain the data on the accelerating expansion of the universe; according to the standard model, the dark energy, which is spread uniformly
throughout the universe, makes up about 70 percent of the total mass and energy content of the universe.

Dark matter—invisible, not-yet-detected, unknown particles of matter, representing about 30 percent of the total mass of matter according to the standard model; its presence is necessary if Newton's and Einstein's gravity theories are to fit data from galaxies, clusters of galaxies, and cosmology; together, dark
matter and dark energy mean that 96 percent of the matter and energy in the universe is invisible.

Deferent—in the ancient Ptolemaic concept of the universe, a large circle representing the orbit of a planet around the Earth.

Doppler principle or **Doppler effect**—the discovery by the nineteenth-century Austrian scientist Christian Doppler that when sound or light waves are moving toward an observer, the apparent frequency of the waves will be shortened, while if they are moving away from an observer, they will be lengthened; in
astronomy this means that the light emitted by galaxies moving away from us is redshifted, and that from nearby galaxies moving toward us is blueshifted.

Dwarf galaxy—a small galaxy (containing several billion stars) orbiting a larger galaxy; the Milky Way has over a dozen dwarf galaxies as companions, including the Large Magellanic Cloud and Small Magellanic Cloud.

Dynamics—the physics of matter in motion.

Electromagnetism—the unified force of electricity and magnetism, discovered to be the same phenomenon by Michael Faraday and James Clerk Maxwell in the nineteenth century.

Electromagnetic radiation—a term for wave motion of electromagnetic fields which propagate with the speed of light—300,000 kilometers per second—and differ only in wavelength; this includes visible light, ultraviolet light, infrared radiation,
X-rays, gamma rays, and radio waves.

Electron—an elementary particle carrying negative charge that orbits the nucleus of an atom.

Eötvös experiments—torsion balance experiments performed by Hungarian Count Roland von Eötvös in the late nineteenth and early twentieth centuries that showed that inertial and gravitational mass were the same to one part in 1011; this was a more accurate determination of the equivalence principle than results achieved by Isaac Newton and, later, Friedrich Wilhelm Bessel.

Epicycle—in the Ptolemaic universe, a pattern of small circles traced out by a planet at the edge of its "deferent" as it orbited the Earth; this was how the Greeks accounted for the apparent retrograde motions of the planets.

Equivalence principle—the phenomenon first noted by Galileo that bodies falling in a gravitational field fall at the same rate, independent of their weight and composition; Einstein extended the principle to show that gravitation is identical (equivalent) to acceleration.

Escape velocity—the speed at which a body must travel in order to escape a strong gravitational field;

rockets fired into orbits around the Earth have calculated escape velocities, as do galaxies at the periphery of galaxy clusters.

Ether (or aether)—a substance whose origins were in the Greek concept of "quintessence," the ether was the medium through which energy and matter moved, something more than a vacuum and less than air; in the late nineteenth century the Michelson-Morley experiment disproved the existence of the ether.

Euclidean geometry—plane geometry developed by the third-century bc Greek mathematician Euclid; in this geometry, parallel lines never meet.

Fermion—a particle with half-integer spin, like protons and electrons, that make up matter.

Field—a physical term describing the forces between massive bodies in gravity and electric charges in electromagnetism; Michael Faraday discovered the concept of field when studying magnetic conductors.

Field equations—differential equations describing the physical properties of interacting massive particles in gravity and electric charges in electromagnetism; Maxwell's equations for electromagnetism and Einstein's equations of gravity are prominent examples in physics.

Fifth force or **"skew" force**—a new force in MOG that has the effect of modifying gravity over limited length scales; it is carried by a particle with mass called the phion.

Fine-tuning—the unnatural cancellation of two or more large numbers involving an absurd number of decimal places, when one is attempting to explain a physical phenomenon; this signals that a true understanding of the physical phenomenon has not been achieved.

Fixed stars—an ancient Greek concept in which all the stars were static in the sky, and moved around the Earth on a distant crystalline sphere.

Frame of reference—the three spatial coordinates and one time coordinate that an observer uses to denote the position of a particle in space and time.

Galaxy—organized group of hundreds of billions of stars, such as our Milky Way.

Galaxy cluster—many galaxies held together by mutual gravity but not in as organized a fashion as stars within a single galaxy.

Galaxy rotation curve—a plot of the Doppler shift data recording the observed velocities of stars in galaxies; those stars at the periphery of giant spiral galaxies are observed to be going faster than they "should be" according to Newton's and Einstein's gravity theories.

General relativity—Einstein's revolutionary gravity theory, created in 1916 from a mathematical generalization of his theory of special relativity; it changed our concept of gravity from Newton's universal force to the warping of the geometry of spacetime in the presence of matter and energy.

Geodesic—the shortest path between two neighboring points, which is a straight line in Euclidian geometry, and a unique curved path in four-dimensional spacetime.

Globular cluster—a relatively small, dense system of up to millions of stars occurring commonly in galaxies.

Gravitational lensing—the bending of light by the curvature of spacetime; galaxies and clusters of galaxies act as lenses, distorting the images of distant bright galaxies or quasars as the light passes through or near them.

Gravitational mass—the active mass of a body that produces a gravitational force on other bodies.

Gravitational waves—ripples in the curvature of spacetime predicted by general relativity; although any accelerating body can produce gravitational radiation or waves, those that could be detected by experiments would be caused by cataclysmic cosmic events.

Graviton—the hypothetical smallest packet of gravitational energy, comparable to the photon for electromagnetic energy; the graviton has not yet been seen experimentally.

Group (in mathematics)—in abstract algebra, a set that obeys a binary operation that satisfies certain axioms; for example, the property of addition of integers makes a group; the branch of mathematics that studies groups is called group theory.

Hadron—a generic word for fermion particles that undergo strong nuclear interactions.

Hamiltonian—an alternative way of deriving the differential equations of motion for a physical system using the calculus of variations; Hamilton's principle is also called the "principle of stationary action" and was originally formulated by Sir William Rowan Hamilton for classical mechanics; the principle applies to classical fields such as the gravitational and electromagnetic fields, and has had important applications in quantum mechanics and quantum field theory.

Homogeneous—in cosmology, when the universe appears the same to all observers, no matter where they are in the universe.

Inertia—the tendency of a body to remain in uniform motion once it is moving, and to stay at rest if it is at rest; Galileo discovered the law of inertia in the early seventeenth century.

Inertial mass—the mass of a body that resists an external force; since Newton, it has been known experimentally that inertial and gravitational mass are equal; Einstein used this equivalence of inertial and gravitational mass to postulate his equivalence principle, which was a cornerstone of his gravity theory.

Inflation theory—a theory proposed by Alan Guth and others to resolve the flatness, horizon, and homogeneity problems in the standard big bang model; the very early universe is pictured as expanding exponentially fast in a fraction of a second.

Interferometry—the use of two or more telescopes, which in combination create a receiver in effect as large as the distance between them; radio astronomy in particular makes use of interferometry.

Inverse square law—discovered by Newton, based on earlier work by Kepler, this law states that the force of gravity between two massive bodies or point particles decreases as the inverse square of the distance between them.

Isotropic—in cosmology, when the universe looks the same to an observer, no matter in which direction she looks.

Kelvin temperature scale—designed by Lord Kelvin (William Thomson) in the mid-1800s to measure very cold temperatures, its starting point is absolute zero, the coldest possible temperature in the universe, corresponding to −273.15 degrees Celsius; water's freezing point is 273.15K (0°C), while its boiling point is 373.15K (100°C).

Lagrange points—discovered by the Italian-French mathematician Joseph-Louis Lagrange, these five special points are in the vicinity of two orbiting masses where a third, smaller mass can orbit at a fixed distance from the larger masses; at the Lagrange points, the gravitational pull of the two large masses precisely equals the centripetal force required to keep the third body, such as a satellite, in a bound orbit; three of the Lagrange points are unstable, two are stable.

Lagrangian—named after Joseph-Louis Lagrange, and denoted by L, this mathematical expression summarizes the dynamical properties of a physical system; it is defined in classical mechanics as the kinetic energy T minus the potential energy V; the equations of motion of a system of particles may be derived from the Euler-Lagrange equations, a family of partial differential equations.

Light cone—a mathematical means of expressing past, present, and future space and time in terms of spacetime geometry; in four-dimensional Minkowski spacetime, the light rays emanating from or arriving at an event separate spacetime into a past cone and a future cone which meet at a point corresponding
to the event.

Lorentz transformations—
mathematical transformations from one inertial frame of reference to another such that the laws of physics remain the same; named after Hendrik Lorentz, who developed them in 1904, these transformations form the basic mathematical equations underlying special relativity.

Mercury anomaly—a phenomenon in which the perihelion of Mercury's orbit advances more rapidly than predicted by Newton's equations of gravity; when Einstein showed that his gravity theory predicted the anomalous precession, it was the first empirical evidence that general relativity might be correct.

Meson—a short-lived boson composed of a quark and an antiquark, believed to bind protons and neutrons together in the atomic nucleus.

Metric tensor—mathematical symmetric tensor coefficients that determine the infinitesimal distance between two points in spacetime; in effect the metric tensor distinguishes between Euclidean and non-Euclidean geometry.

Michelson-Morley experiment—1887 experiment by Albert Michelson and Edward Morley that proved that the ether did not exist; beams of light traveling in the same direction, and in the perpendicular direction, as the supposed ether showed no difference in speed or arrival time at their destination.

Milky Way—the spiral galaxy that contains our solar system.

Minkowski spacetime—the geometrically flat spacetime, with no gravitational effects, first described by the Swiss mathematician Hermann Minkowski; it became the setting of Einstein's theory of gravity.

MOG—my relativistic modified theory of gravitation, which generalizes Einstein's general relativity; MOG stands for "Modified Gravity."

MOND—a modification of Newtonian gravity published by Mordehai Milgrom in 1983; this is a nonrelativistic phenomenological model used to describe rotational velocity curves of galaxies; MOND stands for "Modified Newtonian Dynamics."

Neutrino—an elementary particle with zero electric charge; very difficult to detect, it is created in radioactive decays and is able to pass through matter almost undisturbed; it is considered to have a tiny mass that has not yet been accurately measured.

Neutron—an elementary and electrically neutral particle found in the atomic nucleus, and having about the same mass as the proton.

Nuclear force—another name for the strong force that binds protons and neutrons together in the atomic nucleus.

Nucleon—a generic name for a proton or neutron within the atomic nucleus.

Neutron star—the collapsed core of a star that remains after a supernova explosion; it is extremely dense, relatively small, and composed of neutrons.

Newton's gravitational constant—the constant of proportionality, G, which occurs in the Newtonian law of gravitation, and says that the attractive force between two bodies is proportional to the product of their masses and inversely proportional to the square of the distance between them; its numerical value is: G = 6.67428 ± 0.00067 x 10–11 m3 kg–1 s–2.

Nonsymmetric field theory (Einstein)—a mathematical description of the geometry of spacetime based on a metric tensor that has both a symmetric part and an antisymmetric part; Einstein used this geometry to formulate a unified field

theory of gravitation and electromagnetism.

Nonsymmetric Gravitation Theory (NGT)—my generalization of Einstein's purely gravitation theory (general relativity) that introduces the antisymmetric field as an extra component of the gravitational field; mathematically speaking, the nonsymmetric field structure is described by a non-Riemannian geometry.

Parallax—the apparent movement of a nearer object relative to a distant background when one views the object from two different positions; used with triangulation for measuring distances in astronomy.

Paradigm shift—a revolutionary change in belief, popularized by the philosopher Thomas Kuhn, in which the majority of scientists in a given field discard a traditional theory of nature in favor of a new one that passes the tests of experiment and observation; Darwin's theory of natural selection, Newton's gravity theory, and Einstein's general relativity all represented paradigm shifts.

Parsec—a unit of astronomical length equal to 3.262 light years.

Particle-wave duality—the fact that light in all parts of the electromagnetic spectrum (including radio waves, X-rays, etc., as well as visible light) sometimes acts like waves and sometimes acts like particles or photons; gravitation may be similar, manifesting as waves in spacetime or graviton particles.

Perihelion—the position in a planet's elliptical orbit when it is closest to the sun.

Perihelion advance—the movement, or changes, in the position of a planet's perihelion in successive revolutions of its orbit over time; the most dramatic perihelion advance is Mercury's, whose orbit traces a rosette pattern.

Perturbation theory—a mathematical method for finding an approximate solution to an equation that cannot be solved exactly, by expanding the solution in a series in which each successive term is smaller than the preceding one.

Phion—name given to the massive vector field in MOG; it is represented both by a boson particle, which carries the fifth force, and a field.

Photoelectric effect—the ejection of electrons from a metal by X-rays, which proved the existence of photons; Einstein's explanation of this effect in 1905 won him the Nobel Prize in 1921; separate experiments proving and demonstrating
the existence of photons were performed in 1922 by Thomas Millikan and Arthur Compton, who received the Nobel Prize for this work in 1923 and 1927, respectively.

Photon—the quantum particle that carries the energy of electromagnetic waves; the spin of the photon is 1 times Planck's constant h.

Pioneer 10 and 11 spacecraft—launched by NASA in the early 1970s to explore the outer solar system, these spacecraft show a small, anomalous acceleration as they leave the inner solar system.

Planck's constant (h)—a fundamental constant that plays a crucial role in quantum mechanics, determining the size of quantum packages of energy such as the photon; it is named after Max Planck, a founder of quantum mechanics

Principle of general covariance—Einstein's principle that the laws of physics remain the same whatever the frame of reference an observer uses to measure physical quantities.

Principle of least action—more accurately the principle of *stationary* action, this variational principle, when applied to a mechanical system or a field theory, can be used to derive the equations of motion of the system; the credit for discovering the principle is given to Pierre-Louis Moreau Maupertius but it may have been discovered independently by Leonhard Euler or Gottfried Leibniz.

Proton—an elementary particle that carries positive electrical charge and is the nucleus of a hydrogen atom.

Ptolemaic model of the universe—the predominant theory of the universe until the Renaissance, in which the Earth was the heavy center of the universe and all other heavenly bodies, including the moon, sun, planets, and stars, orbited around it; named for Claudius Ptolemy.

Quantize—to apply the principles of quantum mechanics to the behavior of matter and energy (such as the electromagnetic or gravitational field energy); breaking down a field into its smallest units or packets of energy.

Quantum field theory—the modern relativistic version of quantum mechanics used to describe the physics of elementary particles; it can also be used in nonrelativistic fieldlike systems in condensed matter physics.

Quantum gravity—attempts to unify gravity with quantum mechanics.

Quantum mechanics—the theory of the interaction between quanta (radiation) and matter; the effects of quantum mechanics become observable at the submicroscopic distance scales of atomic and particle physics, but macroscopic quantum effects can be seen in the phenomenon of quantum entanglement.

Quantum spin—the intrinsic quantum angular momentum of an elementary particle; this is in contrast to the classical orbital angular momentum of a body rotating about a point in space.

Quark—the fundamental constituent of all particles that interact through the strong nuclear force; quarks are fractionally charged, and come in several varieties; because they are confined within particles such as protons and neutrons, they cannot be detected as free particles.

Quasars—"quasi-stellar objects," the farthest distant objects that can be detected with radio and optical telescopes; they are exceedingly bright, and are believed to be young, newly forming galaxies; it was the discovery of quasars in 1960 that quashed the steady-state theory of the universe in favor of the big bang.

Quintessence—a fifth element in the ancient Greek worldview, along with earth, water, fire and air, whose purpose was to move the crystalline spheres that supported the heavenly bodies orbiting the Earth; eventually this concept became known as the "ether," which provided the *something* that bodies needed to be in contact with in order to move; although Einstein's special theory of relativity dispensed with the ether, recent explanations of the acceleration of the universe call the varying negative pressure vacuum energy "quintessence."

Redshift—a useful phenomenon based on the Doppler principle that can indicate whether and how fast bodies in the universe are receding from an observer's position on Earth; as galaxies move rapidly away from us, the frequency of the wavelength of their light is shifted toward the red end of the electromagnetic
spectrum; the amount of this shifting indicates the distance of the galaxy.

Riemann curvature tensor—a mathematical term that specifies the curvature of four-dimensional spacetime.

Riemannian geometry—a non-Euclidean geometry developed in the mid-nineteenth century by the German mathematician George Bernhard Riemann that describes curved surfaces on which parallel lines *can* converge, diverge, and even intersect, unlike Euclidean geometry; Einstein made Riemannian geometry the mathematical formalism of general relativity.

Satellite galaxy—a galaxy that orbits a host galaxy or even a cluster of galaxies.

Scalar field—a physical term that associates a value without direction to every point in space, such as temperature, density, and pressure; this is in contrast to a vector field, which has a direction in space; in Newtonian physics or in electrostatics, the potential energy is a scalar field and its gradient is the vector force field; in quantum field theory, a scalar field describes a boson particle with spin zero.

Scale invariance—distribution of objects or patterns such that the same shapes and distributions remain if one increases or decreases the size of the length scales or space in which the objects are observed; a common example of scale invariance
is fractal patterns.

Schwarzschild solution—an exact spherically symmetric static solution of Einstein's field equations in general relativity, worked out by the astronomer Karl Schwarzschild in 1916, which predicted the existence of black holes.

Self-gravitating system—a group of objects or astrophysical bodies held together by mutual gravitation, such as a cluster of galaxies; this is in contrast to a "bound system" like our solar system, in which bodies are mainly attracted to and revolve around a central mass.

Singularity—a place where the solutions of differential equations break down; a spacetime singularity is a position in space where quantities used to determine the gravitational field become infinite; such quantities include the curvature of spacetime and the density of matter.

Spacetime—in relativity theory, a combination of the three dimensions of space with time into a four-dimensional geometry; first introduced into relativity by Hermann Minkowski in 1908.

Special theory of relativity—Einstein's initial theory of relativity, published in 1905, in which he explored the "special" case of transforming the laws of physics from one uniformly moving frame of reference to another; the equations

of special relativity revealed that the speed of light is a constant, that objects appear contracted in the direction of motion when moving at close to the speed of light, and that $E = mc^2$, or energy is equal to mass times the speed of light squared.

Spin—see quantum spin.

String theory—a theory based on the idea that the smallest units of matter are not point particles but vibrating strings; a popular research pursuit in physics for two decades, string theory has some attractive mathematical features, but has yet to make a testable prediction.

Strong force—see nuclear force.

Supernova—spectacular, brilliant death of a star by explosion and the release of heavy elements into space; supernovae type 1a are assumed to have the same intrinsic brightness and are therefore used as standard candles in estimating cosmic distances.

Supersymmetry—a theory developed in the 1970s which, proponents claim, describes the most fundamental spacetime symmetry of particle physics: For every boson particle there is a supersymmetric fermion partner, and for every fermion there exists a supersymmetric boson partner; to date, no supersymmetric particle partner has been detected.

Tully-Fisher law—a relation stating that the asymptotically flat rotational velocity of a star in a galaxy, raised to the fourth power, is proportional to the mass or luminosity of the galaxy.

Unified theory (or unified field theory)—a theory that unites the forces of nature; in Einstein's day those forces consisted of electromagnetism and gravity; today the weak and strong nuclear forces must also be taken into account, and perhaps someday MOG's fifth force or skew force will be included; no one has yet discovered a successful unified theory.

Vacuum—in quantum mechanics, the lowest energy state, which corresponds to the vacuum state of particle physics; the vacuum in modern quantum field theory is the state of perfect balance of creation and annihilation of particles and antiparticles.

Variable Speed of Light (VSL) cosmology—an alternative to inflation theory, in which the speed of light was much faster at the beginning of the universe than it is today; like inflation, this theory solves the horizon and flatness problems in the very early universe in the standard big bang model.

Vector field—a physical value that assigns a field with the position and direction of a vector in space; it describes the force field of gravity or the electric and magnetic force fields in James Clerk Maxwell's field equations.

Virial theorem—a means of estimating the average speed of galaxies within galaxy clusters from their estimated average kinetic and potential energies.

Vulcan—a hypothetical planet predicted by the nineteenth-century astronomer Urbain Jean Joseph Le Verrier to be the closest orbiting planet to the sun; the presence of Vulcan would explain the anomalous precession of the perihelion of Mercury's orbit; Einstein later explained the anomalous precession in general relativity by gravity alone.

Weak force—one of the four fundamental forces of nature, associated with radioactivity such as beta decay in subatomic physics; it is much weaker than the strong nuclear force but still much stronger than gravity.

X-ray clusters—galaxy clusters with large amounts of extremely hot gas within them that emit X-rays; in such clusters, this hot gas represents at least twice the mass of the luminous stars.

Bibliography

Misner, C.W., Thorne, K.S., and Wheeler, J.A. (1973) *Gravitation*, Freeman, San Francisco, p. 5.

Wilson, H.A. (1921) An electromagnetic theory of gravitation, *Phys. Rev.* **17**, 54-59.

Dicke, R.H. (1957) Gravitation without a principle of equivalence, *Rev. Mod. Phys.* **29**, 363-376. See also Dicke, R.H. (1961) Mach's principle and equivalence, in C. Møller (ed.), *Proc. of the Intern'l School of Physics "Enrico Fermi" Course XX, Evidence for Gravitational Theories*, Academic Press, New York, pp.1-49.

Lightman, A.P., and Lee, D.L. (1973) Restricted proof that the weak equivalence principle implies the Einstein equivalence principle, *Phys. Rev.* D **8**, 364-376.

Will, C.M. (1974) Gravitational red-shift measurements as tests of nonmetric theories of gravity, *Phys. Rev.* D **10**, 2330-2337.

Haugan, M.P., and Will, C.M. (1977) Principles of equivalence, Eötvös experiments, and gravitational red-shift experiments: The free fall of electromagnetic systems to post -post-Coulombian order, *Phys. Rev.* D **15**, 2711-2720.

Volkov, A.M., Izmest'ev, A.A., and Skrotskii, G.V. (1971) The propagation of electromagnetic waves in a Riemannian space, *Sov. Phys. JETP* **32**, 686-689.

Heitler, W. (1954) *The Quantum Theory of Radiation*, 3rd ed., Oxford University Press, London, p. 113.

Alpher, R.A. (Jan.-Feb. 1973) Large numbers, cosmology, and Gamow, *Am. Sci.* **61**, 52-58.

Harrison, E.R. (Dec. 1972) The cosmic numbers, *Phys. Today* **25**, 30-34.

Webb, J.K., Flambaum, V.V., Churchill, C.W., Drinkwater, M.J., and Barrow, J.D. (1999) Search for time variation of the fine structure constant, *Phys. Rev. Lett.* **82**, 884-887.

Brault, J.W. (1963) Gravitational red shift of solar lines, *Bull. Amer. Phys. Soc.* **8**, 28.

Pound, R.V., and Rebka, G.A. (1960) Apparent weight of photons, *Phys. Rev. Lett.* **4**, 337-341.

Pound, R.V., and Snider, J.L. (1965) Effect of gravity on nuclear resonance, *Phys. Rev. Lett.* **13**, 539-540.

Goldstein, H. (1957) *Classical Mechanics*, Addison-Wesley, Reading MA, pp. 206-207.

Mizobuchi, Y. (1985) New theory of space-time and gravitation - Yilmaz's approach, *Hadronic Jour.* **8**, 193-219.

Alley, C.O. (1995) The Yilmaz theory of gravity and its compatibility with quantum theory, in D.M. Greenberger and A. Zeilinger (eds.), *Fundamental Problems in Quantum Theory: A Conference Held in Honor of Professor John A. Wheeler*, Vol. 755 of the Annals of the New York Academy of Sciences, New York, pp. 464-475.

Schilling, G.(1999) Watching the universe's second biggest bang, *Science* **283**, 2003-2004.

Robertson, S.L. (1999) Bigger bursts from merging neutron stars, *Astrophys. Jour.* **517**, L117-L119.

Hughes, V.W., Robinson, H.G. and Beltran-Lopez, V. (1960) Upper limit for the anisotropy of inertial mass from nuclear resonance experiments, *Phys. Rev. Lett.* **4**, 342-344.

Drever, R.W.P. (1961) A search for anisotropy of inertial mass using a free precession technique, *Phil. Mag.* **6**, 683-687.

Collela, R. Overhauser, A.W., and Werner, S.A. (1975) Observation of gravitationally induced quantum mechanics, *Phys. Rev Lett.* **34**, 1472-1474.

Puthoff, H.E. (1996) SETI, the velocity-of-light limitation, and the Alcubierre warp drive: an integrating overview, *Physics Essays* **9**, 156-158.

Atkinson, R. d'E. (1962) General relativity in Euclidean terms, *Proc. Roy. Soc.* **272**, 60-7

Barton, G.; Scharnhorst, K. (1993). "QED Between Parallel Mirrors: Light Signals Faster Than c, or Amplified by the Vacuum". Journal of Physics A: Mathematical and General. **26** (8): 2037–2046. *Bibcode:1993JPhA...26.2037B. doi:10.1088/0305-4470/26/8/024. ISSN 0305-4470*.

Beiser, A. (2003). *Concepts of Modern Physics* (6th ed.). Boston: McGraw-Hill. *ISBN 978-0072448481. LCCN 2001044743. OCLC 48965418*.

Bordag, M; Klimchitskaya, G. L.; Mohideen, U.; Mostepanenko, V. M. (2009). *Advances in the Casimir Effect*. Oxford: Oxford University Press. ISBN *978-0-19-923874-3*. LCCN *2009279136*. OCLC *319209483*.

Boyer, T. H. (1970). "Quantum Zero-Point Energy and Long-Range Forces". Annals of Physics. 56 (2): 474–503. Bibcode:*1970AnPhy..56..474B*. doi:*10.1016/0003-4916(70)90027-8*. ISSN *0003-4916*. OCLC *4648258537*.

Carroll, S. M.; Field, G. B. (1997). *"Is There Evidence for Cosmic Anisotropy in the Polarization of Distant Radio Sources?"* (PDF). Physical Review Letters. 79 (13): 2394–2397. arXiv:*astro-ph/9704263*.

Conlon, T. E. (2011). *Thinking About Nothing: Otto Von Guericke and The Magdeburg Experiments on the Vacuum*. San Francisco: Saint Austin Press. ISBN *978-1-4478-3916-3*. OCLC *840927124*.

Davies, P. C. W. (1985). *Superforce: The Search for a Grand Unified Theory of Nature*. New York: Simon and Schuster. ISBN *978-0-671-47685-4*. LCCN *84005473*. OCLC *12397205*.

Dunne, G. V. (2012). "The Heisenberg-Euler Effective Action: 75 years on". International Journal of Modern Physics A. 27 (15): 1260004. arXiv:*1202.1557*. Bibcode:*2012IJMPA..2760004D*. doi:*10.1142/S0217751X12600044*. ISSN *0217-751X*.

Einstein, A. (1995). Klein, Martin J.; Kox, A. J.; Renn, Jürgen; Schulmann, Robert (eds.). *The Collected Papers of Albert Einstein Vol. 4 The Swiss Years: Writings, 1912–1914*.

Princeton: Princeton University Press. ISBN 978-0-691-03705-9. OCLC 929349643.

Greiner, W.; Müller, B.; Rafelski, J. (2012). *Quantum Electrodynamics of Strong Fields: With an Introduction into Modern Relativistic Quantum Mechanics*. Springer. doi:10.1007/978-3-642-82272-8. ISBN 978-0-387-13404-8. LCCN 84026824. OCLC 317097176.

Haisch, B.; Rueda, A.; Puthoff, H. E. (1994). *"Inertia as a Zero-Point-Field Lorentz Force"* (PDF). Physical Review A. 49 (2): 678–694. Bibcode:1994PhRvA..49..678H. doi:10.1103/PhysRevA.49.678. PMID 9910287.

Heisenberg, W.; Euler, H. (1936). "Folgerungen aus der Diracschen Theorie des Positrons". Zeitschrift für Physik. 98 (11–12): 714–732. arXiv:physics/0605038. Bibcode:1936ZPhy...98..714H. doi:10.1007/BF01343663. ISSN 1434-6001.

Heitler, W. (1984). *The Quantum Theory of Radiation* (1954 reprint 3rd ed.). New York: Dover Publications. ISBN 978-0486645582. LCCN 83005201. OCLC 924845769.

Heyl, J. S.; Shaviv, N. J. (2000). "Polarization evolution in strong magnetic fields". Monthly Notices of the Royal Astronomical Society. 311 (3): 555–564. arXiv:astro-ph/9909339. Bibcode:2000MNRAS.311..555H. doi:10.1046/j.1365-8711.2000.03076.x. ISSN 0035-8711.

Itzykson, C.; Zuber, J.-B. (1980). *Quantum Field Theory* (2005 ed.). Mineola, New York: Dover Publications. ISBN 978-0486445687. LCCN 2005053026. OCLC 61200849.

Kostelecký, V. Alan; Mewes, M. (2009). "Electrodynamics with Lorentz-violating operators of arbitrary dimension". Physical Review D. 80 (1): 015020. *arXiv:0905.0031*. *Bibcode:2009PhRvD..80a5020K*. *doi:10.1103/PhysRevD.80.015020*. *ISSN 1550-7998*.

Kostelecký, V. Alan; Mewes, M. (2013). "Constraints on Relativity Violations from Gamma-Ray Bursts". Physical Review Letters. 110 (20): 201601. *arXiv:1301.5367*. *Bibcode:2013PhRvL.110t1601K*. *doi:10.1103/PhysRevLett.110.201601*. *ISSN 0031-9007*. *PMID 25167393*.

Kragh, H. (2012). "Preludes to Dark Energy: Zero-Point Energy and Vacuum Speculations". Archive for History of Exact Sciences. 66 (3): 199–240. *arXiv:1111.4623*. *doi:10.1007/s00407-011-0092-3*. *ISSN 0003-9519*.

Kragh, H. S.; Overduin, J. M. (2014). *The Weight of the Vacuum: A Scientific History of Dark Energy*. New York: Springer. *ISBN 978-3-642-55089-8*. *LCCN 2014938218*. *OCLC 884863929*.

Kuhn, T. (1978). *Black-Body Theory and the Quantum Discontinuity, 1894-1912*. New York: Oxford University Press. *ISBN 978-0-19-502383-1*. *LCCN 77019022*. *OCLC 803538583*.

Lahteenmaki, P.; Paraoanu, G. S.; Hassel, J.; Hakonen, P. J. (2013). "Dynamical Casimir Effect in a Josephson Metamaterial". Proceedings of the National Academy of Sciences. 110 (11): 4234–4238. *arXiv:1111.5608*. *Bibcode:2013PNAS..110.4234L*. *doi:10.1073/pnas.1212705110*. *ISSN 0027-8424*.

Leuchs, G.; Sánchez-Soto, L. L. (2013). "A Sum Rule For Charged Elementary Particles". The European Physical Journal D. 67 (3): 57. *arXiv*:*1301.3923*. Bibcode:*2013EPJD...67...57L*. doi:*10.1140/epjd/e2013-30577-8*. ISSN *1434-6060*.

Loudon, R. (2000). *The Quantum Theory of Light* (3rd ed.). Oxford: Oxford University Press. ISBN *978-0198501770*. LCCN *2001265846*. OCLC *44602993*.

Mignani, R. P.; Testa, V.; González Caniulef, D.; Taverna, R.; Turolla, R.; Zane, S.; Wu, K. (2017). *"Evidence for vacuum birefringence from the first optical-polarimetry measurement of the isolated neutron star RX J1856.5−3754"* (PDF). Monthly Notices of the Royal Astronomical Society. 465 (1): 492–500. *arXiv*:*1610.08323*. Bibcode:*2017MNRAS.465..492M*. doi:*10.1093/mnras/stw2798*. ISSN *0035-8711*.

Milonni, P. W. (1994). *The Quantum Vacuum: An Introduction to Quantum Electrodynamics*. Boston: Academic Press. ISBN *978-0124980808*. LCCN *93029780*. OCLC *422797902*.
Milonni, P. W. (2009). *"Zero-Point Energy"*. In Greenberger, Daniel; Hentschel, Klaus; Weinert, Friedel (eds.). Compendium of Quantum Physics: Concepts, Experiments, History and Philosophy. In D. Greenberger, K. Hentschel and F. Wienert (Eds.), Compendium of Quantum Physics Concepts, Experiments, History and Philosophy (Pp.). Berlin, Heidelberg: Springer. pp. 864–866. *arXiv*:*0811.2516*. doi:*10.1007/978-3-540-70626-7*. ISBN *9783540706229*. LCCN *2008942038*. OCLC *297803628*.

Peebles, P. J. E.; *Ratra, Bharat* (2003). "The Cosmological Constant and Dark Energy". Reviews of Modern Physics. 75 (2): 559–606. *arXiv:astro-ph/0207347*. *Bibcode:2003RvMP...75..559P*. *doi:10.1103/RevModPhys.75.559*. *ISSN 0034-6861*.

Power, E. A. (1964). Introductory Quantum Electrodynamics. London: Longmans. *LCCN 65020006*. *OCLC 490279969*.

Rafelski, J.; Muller, B. (1985). *Structured Vacuum: Thinking About Nothing* (PDF). H. Deutsch: Thun. *ISBN 978-3871448898*. *LCCN 86175968*. *OCLC 946050522*.

Rees, Martin, ed. (2012). *Universe*. New York: DK Pub. *ISBN 978-0-7566-9841-6*. *LCCN 2011277855*. *OCLC 851193468*.

Riek, C.; Seletskiy, D. V.; Moskalenko, A. S.; Schmidt, J. F.; Krauspe, P.; Eckart, S.; Eggert, S.; Burkard, G.; Leitenstorfer, A. (2015). *"Direct Sampling of Electric-Field Vacuum Fluctuations"* (PDF). Science. 350 (6259): 420–423. *Bibcode:2015Sci...350..420R*. *doi:10.1126/science.aac9788*. *ISSN 0036-8075*. *PMID 26429882*.

Rugh, S. E.; Zinkernagel, H. (2002). "The Quantum Vacuum and the Cosmological Constant Problem". Studies in History and Philosophy of Science Part B: Studies in History and Philosophy of Modern Physics. 33 (4): 663–705. *arXiv:hep-th/0012253*. *Bibcode:2002SHPMP..33..663R*. *doi:10.1016/S1355-2198(02)00033-3*. *ISSN 1355-2198*.

Schwinger, J. (1998a). Particles, Sources, and Fields: Volume I. Reading, Massachusetts: Advanced Book Program,

Perseus Books. *ISBN* *978-0-7382-0053-8*. *LCCN* *98087896*. *OCLC* *40544377*.

Schwinger, J. (1998b). Particles, Sources, and Fields: Volume II. Reading, Massachusetts: Advanced Book Program, Perseus Books. *ISBN* *978-0-7382-0054-5*. *LCCN* *98087896*. *OCLC* *40544377*.

Schwinger, J. (1998c). Particles, Sources, and Fields: Volume III. Reading, Massachusetts: Advanced Book Program, Perseus Books. *ISBN* *978-0-7382-0055-2*. *LCCN* *98087896*. *OCLC* *40544377*.

Sciama, D. W. (1991). "The Physical Significance of the Vacuum State of a Quantum Field". In *Saunders, Simon*; *Brown, Harvey R.* (eds.). The Philosophy of Vacuum. Oxford: Oxford University Press. *ISBN* *978-0198244493*. *LCCN* *90048906*. *OCLC* *774073198*.

Saunders, Simon; *Brown, Harvey R.*, eds. (1991). The Philosophy of Vacuum. Oxford: Oxford University Press. *ISBN* *978-0198244493*. *LCCN* *90048906*. *OCLC* *774073198*.

Urban, M.; Couchot, F.; Sarazin, X.; Djannati-Atai, A. (2013). "The Quantum Vacuum as the Origin of the Speed of Light". The European Physical Journal D. 67 (3): 58. *arXiv*:*1302.6165*. *Bibcode*:*2013EPJD...67...58U*. *doi*:*10.1140/epjd/e2013-30578-7*. *ISSN* *1434-6060*.

Weinberg, S. (1989). *"The Cosmological Constant Problem"* *(PDF)*. Reviews of Modern Physics. 61 (1): 1–23. *Bibcode*:*1989RvMP...61....1W*. *doi*:*10.1103/RevModPhys.61.1*. *hdl*:*2152/61094*. *ISSN* *0034-6861*.

Weinberg, S. (2015). Lectures on Quantum Mechanics (2nd ed.). Cambridge: Cambridge University Press. ISBN 978-1-107-11166-0. LCCN 2015021123. OCLC 910664598.

Weisskopf, V. (1936). *"Über die Elektrodynamik des Vakuums auf Grund des Quanten-Theorie des Elektrons"* (PDF). Kongelige Danske Videnskabernes Selskab, Mathematisk-fysiske Meddelelse. 24 (6): 3–39.

Wilson, C. M.; Johansson, G.; Pourkabirian, A.; Simoen, M.; Johansson, J. R.; Duty, T.; Nori, F.; Delsing, P. (2011). "Observation of the Dynamical Casimir Effect in a Superconducting Circuit". Nature. 479 (7373): 376–379. arXiv:1105.4714. Bibcode:2011Natur.479..376W. doi:10.1038/nature10561. ISSN 0028-0836. PMID 22094697.

Balungi Francis, (2010) "A hypothetical investigation into the realm of the microscopic and macroscopic universes beyond the standard model" general physics arXiv:1002.2287v1 [physics.gen-ph]

Hawking, Stephen (1975). "Particle Creation by Black Holes". Commun. Math. Phys. 43 (3): 199–220. Bibcode:1975CMaPh..43..199H.

Hawking, S. W. (1974). "Black hole explosions?". Nature.248(5443):30–31. Bibcode:1974Natur.248...30H.doi:10.1038/248030a0.

Carlo Rovelli (2003) "Quantum Gravity" Draft of the Book Pdf
Some few texts used are from Wikipedia https://creativecommons.org/licenses/by-sa/3.0/
D. N. Page, Phys. Rev. D 13, 198 (1976).

C. Gao and Y.Lu, Pulsations of a black hole in LQG (2012) arXiv:1706.08009v3

A.H. Chamseddine and V.Mukhanov, Non singular black hole (2016) arXiv 1612.05861v1

M.Bojowald and G.M.Paily, A no-singularity scenario in LQG (2012) arXiv: 1206.5765v1

P.Singh, class.Quant.Grav,26,125005(2009), arXiv:0901.2750

P.Singh and F.Vidotto, Phys.Rev, D83,064027(2011) arXiv:1012.1307

C.Rovelli and F.Vidotto, Phy. Rev,111(9) 091303(2013) arXiv:1307.3228v2

M.Bojowald, Initial conditions for a universe, Gravity Research Foundation (2003)

A.Ashtekar, Singularity Resolution in Loop Quantum Cosmology (2008) arXiv:0812.4703v1

J.Brunneumann and T.Thiemann, On singularity avoidance in Loop Quantum Gravity (2005) arXiv:0505032v1

L.Modesto, Disappearence of the Black hole singularity in Quantum gravity (2004) arXiv:0407097v2

Mikhailov, A.A. (1959).Mon. Not. Roy. Astron. Soc.,119, 593.

P. Merat etal.(1974). Astron & Astrophys 32, 471-475

Trempler, R.J. (1956).Helv. Phys. Acta, Suppl.,IV, 106.

Trempler, R.J. (1932). " The deflection of light in the sun's gravitational field "Astronomical society of the pacific 167

Einstein, A. (1916).Ann. d. Phys.,49, 769; (1923).The Principle of Relativity, (translators Perret, W. and Jeffery, G.B.), (Dover Publications, Inc., New York), pp. 109–164.

Von Klüber, H. (1960). InVistas in Astronomy, Vol. 3, pp. 47–77.

K. Hentschel (1992). Erwin Finlay Freundlich and testing Einstein theory of relativity, Communicated by J.D. North Muhleman, D.O., Ekers, R.D. and Fomalont, E.B. (1970).Phys. Rev. Lett.,24, 1377

Mikhailov, A.A. (1956).Astron. Zh.,33, 912.

Dyson, F.W., Eddington, A.S. and Davidson, C. (1920).Phil. Trans. Roy. sog., A220, 291
Chant, C.A. and Young, R.K. (1924).Publ. Dom. Astron. Obs.,2, 275.

Campbell, W.W. and Trumbler, R.J. (1923).Lick Obs. Bull.,11, 41.

Freundlich, E.F., von Klüber, H. and von Brunn, A. (1931).Abhandl. Preuss. Akad. Wiss. Berlin, Phys. Math. Kl., No.l;Z. Astrophys.,3, 171

Mikhailov, A.A. (1949).Expeditions to Observe the Total Solar Eclipse of 21 September, 1941, (report), (ed.

Fesenkov, V.G.), (Publications of the Academy of Sciences, U.S.S.R.), pp. 337–351.

S.P. Martin, in Perspectives on Supersymmetry, edited by G.L. Kane (World Scientific, Singapore, 1998) pp. 1–98; and a longer archive version in hep-ph/9709356; I.J.R. Aitchison, hep-ph/0505105.

Mara Beller, Quantum Dialogue: The Making of a Revolution. University of Chicago Press, Chicago, 2001.

Morrison, Philp: "The Neutrino, scientific American, Vol 194,no.1 (1956),pp.58-68.
R. Haag, J. T. Lopuszanski and M. Sohnius, Nucl. Phys. B88, 257 (1975) S.R. Coleman and J. Mandula, Phys.Rev. 159 (1967) 1251.

H.P. Nilles, Phys. Reports 110, 1 (1984).

P. Nath, R. Arnowitt, and A.H. Chamseddine, Applied N = 1 Supergravity (World Scientific, Singapore, 1984).

S.P. Martin, in Perspectives on Supersymmetry, edited by G.L. Kane (World Scientific, Singapore, 1998) pp. 1–98; and a longer archive version in hep-ph/9709356; I.J.R. Aitchison, hep-ph/0505105.

S. Weinberg, The Quantum Theory of Fields, VolumeIII: Supersymmetry (Cambridge University Press, Cambridge,UK, 2000).

E. Witten, Nucl. Phys. B188, 513 (1981).

S. Dimopoulos and H. Georgi, Nucl. Phys. B193, 150(1981).

N. Sakai, Z. Phys. C11, 153 (1981);R.K. Kaul, Phys. Lett. 109B, 19 (1982).

L. Susskind, Phys. Reports 104, 181 (1984).
L. Girardello and M. Grisaru, Nucl. Phys. B194, 65(1982); L.J. Hall and L. Randall,

Phys. Rev. Lett. 65, 2939(1990);I. Jack and D.R.T. Jones, Phys. Lett. B457, 101 (1999).

For a review, see N. Polonsky, Supersymmetry: Structureand phenomena. Extensions of the standard model, Lect.Notes Phys. M68, 1 (2001).

G. Bertone, D. Hooper and J. Silk, Phys. Reports 405, 279 (2005).

G. Jungman, M. Kamionkowski, and K. Griest, Phys. Reports 267, 195 (1996).

V. Agrawal, S.M. Barr, J.F. Donoghue and D. Seckel,Phys. Rev. D57, 5480 (1998).

N. Arkani-Hamed and S. Dimopoulos, JHEP 0506, 073(2005); G.F. Giudice and A. Romanino, Nucl. Phys. B699, 65(2004) [erratum: B706, 65 (2005)]. July 27, 2006 11:28

en.wikipedia.org/wiki/Supersymmetry - 52k - Cached - Similar pages
en.wikipedia.org/wiki/Grand_unification_theory - 39k - Cached - Similar pages

In cosmology, the Planck epoch (or Planck era), named after Max Planck, is the earliest period of time in the history of the universe, en.wikipedia.org/wiki/**Planck_epoch** - 23k - Cached - Similar pages

L. Shapiro and J. Sol`a, Phys. Lett. B 530, 10 (2002);

E. V.Gorbar and I. L. Shapiro, JHEP 02, 021 (2003); A. M. Pelinson,

L. Shapiro, and F. I. Takakura, Nucl. Phys. B 648, 417 (2003).

A. Starobinsky, Phys. Lett. B 91, 99 (1980).

G. F. R. Ellis, J. Murugan, and C. G. Tsagas, Class. Quant. Grav.21, 233 (2004).

H. V. Peiris et al., Astrophys. J. Suppl. 148, 213 (2003).

D. N. Spergel et al., astro-ph/0603449.

Vilenkin, Phys. Rev. D 32, 2511 (1985).

A. Starobinsky, Pis'ma Astron. Zh 9, 579 (1983).

A.H. Guth, Phys. Rev. D23, 347 (1981).

A.D. Linde, Phys. Lett. B108, 389 (1982); A. Albrecht, P.J.

Steinhardt, Phys.Rev. Lett. 48, 1220 (1982).

A.D. Linde, Phys Lett. B129, 177 (1983).

N. Makino, M. Sasaki, Prog. Theor. Phys. 86, 103 (1991);

D. Kaiser, Phys. Rev.D52, 4295 (1995).

H. Goldberg, Phys. Rev. Lett. 50, 1419 (1983).

E. Kolb and M. Turner, The Early Universe (Addison-Wesley, Reading, MA,1990).

W. Garretson and E. Carlson, Phys. Lett. B 315, 232(1993); H. Goldberg, hep-ph/0003197.

Eddington, A. S., The Internal Constitution of the Stars (Cambridge University Press, England,1926), p. 16

Duncan R.C. & Thompson C., Ap.J.392, L 9 (1992).
Thompson, C, Duncan, R.C., Woods, P., Kouveliotou, C., Finger, M.H. & van Parad ij s, J.,ApJ, submitted, astro-ph /9908086, (2000).

Schwinger, J.,Phys. Rev.73, 416L (1948)

Carlip, S.: Quantum gravity: a progress report. Rept. Prog. Phys. 64, 885 (2001).arXiv:gr-qc/0108040

Kerr,R.P.: Gravitational field of a spinning mass as an example of algebraically special metrics.

Phys. Rev. Lett. 11, 237–238 (1963)

Bekenstein, J.: Black holes and the second law. Lett. Nuovo Cim. 4, 737–740 (1972)

Bardeen, J.M., Carter, B., Hawking, S.: The four laws of black hole mechanics. Commun.

Math. Phys. 31, 161–170 (1973)

Tolman, R.: Relativity, Thermodynamics, and Cosmology. Dover Books on Physics Series.

Dover Publications, New York (1987)
Oppenheimer, J., Volkoff, G.: On massive neutron cores. Phys. Rev. 55, 374–381 (1939)

Tolman, R.C.: Static solutions of einstein's field equations for spheres of fluid, Phys. Rev. 55,364–373 (1939)

Zel'dovich Y.B.: Zh. Eksp. Teoret. Fiz.41, 1609 (1961)

Bondi, H.: Massive spheres in general relativity. Proc. Roy. Soc. Lond. A281, 303–317 (1964)

Sorkin, R.D., Wald, R.M., Zhang, Z.J.: Entropy of selfgravitating radiation. Gen. Rel. Grav. 1127–1146 (1981)

Newman, E.T., Couch, R., Chinnapared, K., Exton, A., Prakash, A., et al.: Metric of a rotating,charged mass. J. Math. Phys. 6, 918–919 (1965)

Ginzburg, V., Ozernoi, L.: Sov. Phys. JETP 20, 689 (1965)

Doroshkevich, A., Zel'dovich, Y., Novikov I.: Gravitational collapse of nonsymmetric and rotating masses, JETP 49 (1965)

Israel, W.: Event horizons in static vacuum space-times. Phys. Rev. 164, 1776–1779 (1967)
Israel,W.: Event horizons in static electrovac space-times. Commun. Math. Phys. 8, 245–260 (1968)

Loop quantum gravity does provide such a prediction [363, 364], and it disagrees with the semiclassical

Carter, B.: Axisymmetric black hole has only two degrees of freedom. Phys. Rev. Lett. 26, 331–333 (1971)

Penrose, R.: Gravitational collapse: the role of general relativity. Riv. Nuovo Cim. 1, 252–276 (1969)

Christodoulou, D.: Reversible and irreversible transformations in black hole physics. Phys. Rev. Lett. 25, 1596–1597 (1970)

Christodoulou, D., Ruffini, R.: Reversible transformations of a charged black hole. Phys. Rev. D4, 3552–3555 (1971)

Hawking, S.: Particle creation by black holes. Commun. Math. Phys. 43, 199–220 (1975)

Klein, O.: Die reflexion von elektronen an einem potential sprung nach der relativistischen dynamik von dirac. Z. Phys. 53, 157 (1929)

Wald, R.M.: General Relativity. The University of Chicago Press, Chicago (1984)

Hawking, S.W.: Black hole explosions. Nature 248, 30–31 (1974)

Hawking, S., Ellis, G.: The large scale structure of space-time. Cambridge University Press, Cambridge (1973)

Carter, B.: Black hole equilibrium states, In Black Holes—Les astres occlus. Gordon and Breach Science Publishers, (1973)

Hawking, S.W.: Gravitational radiation from colliding black holes. Phys. Rev. Lett. 26, 1344– 1346 (1971)

Hawking, S.: Black holes in general relativity. Commun. Math. Phys. 25, 152–166 (1972)

Bekenstein, J.: Extraction of energy and charge from a black hole. Phys. Rev. D7, 949–953 (1973)

Bekenstein, J.D.: Black holes and entropy. Phys. Rev. D7, 2333–2346 (1973)

Hawking, S.: Quantum gravity and path integrals. Phys. Rev. D18, 1747–1753 (1978)
Gross, D.J., Perry, M.J., Yaffe, L.G.: Instability of flat space at finite temperature. Phys. Rev. D25, 330–355 (1982)

Unruh, W.G., Wald, R.M.: What happens when an accelerating observer detects a rindler particle. Phys. Rev. D29, 1047–1056 (1984)

Bekenstein, J.D.: A universal upper bound on the entropy to energy ratio for bounded systems. Phys. Rev. D23, 287 (1981)

Unruh,W.,Wald, R.M.: Acceleration radiation and generalized second law of thermodynamics. Phys. Rev. D25, 942–958 (1982)

Unruh, W., Wald, R.M.: Entropy bounds, acceleration radiation, and the generalized second law. Phys. Rev. D27, 2271–2276 (1983)

Image : MPI for gravitational physics / W.Benger-zib

Tomilin,K.A., (1999). "Natural Systems Of Units: To The Centenary Aniniversary Of   The Planck Systems", 287-296

Sivaram, C. (2007). "What Is Special About the Planck Mass"? arXiv:0707.0058v1

H. Georgi and S.L. Glahow. (1974) "Unity Of All Elementary-Particle Forces". Phys. Rev. Letters 32, 438

Luigi Maxmilian Caligiuri, Amrit Sorli. Gravity Originates from Variable Energy Density of Quantum Vacuum. American Journal of Modern Physics. Vol. 3, No. 3, 2014, pp. 118-128. doi: 10.11648/j.ajmp.20140303.11

Philip J. Tattersall,(2018) Quantum Vacuum Energy and the Emergence of Gravity. doi:10.5539/apr.v10n2p1

H. E. Puthoff (1989) Gravity as a zero-point-fluctuation force PHYSICAL REVIEW A VOLUME 39, NUMBER 5

Balungi Francis (2018) "Quantum Gravity in a Nutshell1" Book.

E.Verlinde (2016) Emergent Gravity and the Dark Universe, arXiv:1611.02269v2[hep-th]

S.Hossenfelder (2018) The Redshift-Dependence of Radial Acceleration: Modified gravity versus particle dark matter, arXiv:1803.08683v1[gr-qc]

Robert J. Scherrer (2004) Purely kinetic k-essence as unified dark matter, arXiv:astro-ph/0402316v3

J.S.Farnes (2018), A unifying theory of dark energy and dark matter: Negative masses and matter creation within a modified ΛCDM framework, arXiv:1712.07962v2[physics.gen-ph]

Gustav M Obermair (2013), Primordial Planck mass black holes (PPMBHs) as candidates for dark matter? J. Phys:conf.Ser.442012066

V.Cooray etal...(2017), An alternative approach to estimate the vacuum energy density of free space, doi:10.20944/preprints201707.0048.v1

M.Milgrom, (1983) A modification of the Newtonian dynamics: Implications for galaxies, Astrophys.J.270, 371.

Acknowledgments

This book would never have been completed without the patience and dedication of my wife, Wanyana Ritah. She performed the wonderful and difficult task of editing major parts of the book and helped in researching many details necessary to complete it.

I wish to thank several colleagues for their help and extensive comments on the manuscript. I particularly thank a total of 200 online physics friends and SUSP science foundation members, for a careful reading of the manuscript. Many graduate students have contributed over the years to developing my Quantum theory of gravity.

I also wish to thank my editors, at SUSP science Foundation and Bill stone Services for their enthusiasm and support. Finally, I thank our family for their patience, love, and support during the four years of working on this book.

About the Author

Balungi Francis was born in Kampala, Uganda, to a single poor mother, grew up in Kawempe, and later joined Makerere Universty in 2006, graduating with a Bachelor Science degree in Land Surveying in 2010. For four years he taught in Kampala City high schools, majoring in the fields of Gravitation and Quantum Physics. His first book, "Mathematical Foundation of the Quantum theory of Gravity," won the Young Kampala Innovative Prize and was mentioned in the African Next Einstein Book Prize (ANE).

He has spent over 15years researching and discovering connections in physics, mathematics, geometry, cosmology, quantum mechanics, gravity, in addition to astrophysics, unified physics and geographical information systems . These studies led to his groundbreaking theories, published papers, books and patented inventions in the science of Quantum Gravity, which have received worldwide recognition.

From these discoveries, Balungi founded the SUSP (Solutions to the Unsolved Scientific Problems) Project Foundation in 2004 - now known as the SUSP Science Foundation. As its current Director of Research, Balungi leads physicists, mathematicians and engineers in exploring Quantum Gravity principles and their implications in our world today and for future generations.

Balungi launched the Visionary School of Quantum Gravity in 2016 in order to bring the learning and community further together. It's the first and only

Quantum Gravity physics program of its kind, educating thousands of students from over 80 countries.

The book "Quantum Gravity in a Nutshell1", a most recommend book in quantum gravity research, was produced based on Balungi's discoveries and their potential for generations to come. Balungi is currently guiding the Foundation, speaking to audiences worldwide, and continuing his groundbreaking research.

Contact Balungi Francis at the following address:

Email: balungif@gmail.com
bfrancis@cedat.mak.ac.ug

BILL STONE SERVICES

Tel: +256703683756
+256777105605

Ingram Content Group UK Ltd.
Milton Keynes UK
UKHW010131060623
422929UK00004B/130